Planned Maintenance
for Productivity and
Energy Conservation

SECOND EDITION

Planned Maintenance for Productivity and Energy Conservation

SECOND EDITION

John W. Criswell

Published by
THE FAIRMONT PRESS, INC.
700 Indian Trail
Lilburn, GA 30247

Library of Congress Cataloging-in-Publication Data

Criswell, John W., 1928-
  Planned maintenance for productivity and energy conservation.

  Includes index.
  1. Plant maintenance. 2. Plant maintenance—Forms.
I. Title.
TS192.C74   1987   658.2'02   86-46137
ISBN 0-88173-029-7

Planned Maintenance for
Productivity and Energy Conservation

Second Edition

Published by The Fairmont Press, Inc.
700 Indian Trail
Lilburn, GA 30247

ISBN 0-88173-029-7   FP

ISBN 0-13-679549-8   PH

While every effort is made to provide dependable information, the publisher, author, and editors cannot be held responsible for any errors or omissions.

Printed in the United States of America

Distributed by Prentice-Hall, Inc.
A division of Simon & Schuster
Englewood Cliffs, NJ 07632

Prentice-Hall International (UK) Limited, *London*
Prentice-Hall of Australia Pty. Limited, *Sydney*
Editora Prentice-Hall do Brasil, Ltda., *Rio de Janeiro*
Prentice-Hall Canada Inc., *Toronto*
Prentice-Hall Hispanoamericana, S.A., *Mexico*
Prentice-Hall of India Private Limited, *New Delhi*
Prentice-Hall of Japan, Inc., *Tokyo*
Prentice-Hall of Southeast Asia Pte. Ltd., *Singapore*

Contents

Preface

Planned maintenance is essential for companies to stay competitive. It is the sure way to reduce utility costs and achieve the highest rate of return on investment. *Planned Maintenance for Productivity Energy Conservation* is a practical step-by-step guide presented in forthright terms. It is aimed at the maintenance person who is responsible to get the job done. Mr. Criswell has provided invaluable guidance on ways to improve any maintenance program.

The second edition contains a comprehensive chapter on how to computerize the maintenance function. Included is an extensive guide on microcomputer application software.

Mr. Criswell's sound advice will help any facility, large or small, increase its productivity and slash operating costs.

Albert Thumann, P.E., C.E.M.

The Rationale

Although American institutions, hotels, housing complexes, hospitals, and industry are operating at high levels in planning, they leave much to be desired in efficiency and production. Physical plant operations, and especially maintenance, are still, in most cases, at a pre-World War II stage. The low efficiency and profitability, compounded of late by sky-rocketing fuel costs, have finally brought the need for efficiency of equipment and facility use to the prominence it deserves. Billions of dollars are wasted every year through lack of a functioning maintenance program and in-house energy management programs. Breakdown maintenance and repair programs not only waste the time, energy, and money of those who own and operate facilities and equipment but also—and far more seriously—cause loss of use (production), waste of new product materials, inflation of production personnel salaries due to downtime, ruined capital equipment, and lack of finished product.

The directions, instructions, recommendations, and forms in this program can save time, energy, and money in every physical plant, regardless of its size. The direct, concise explanations and procedures are easy to follow and to apply. A ten-week installation period can turn an unreliable, costly maintenance operation into a functioning, reliable, efficient and profitable one. Reports that are totally lacking in pertinent information will be replaced by forms and reports containing correct concise information and revealing facts. In addition, the program provides the organization with an opportunity to apply the general principles to its own particular circumstances and then, using the data gathered, to achieve effectiveness and savings in its operations.

About the Author

JOHN W. CRISWELL has spent most of his life involved in maintenance, instruction, management, and energy conservation. He served in the U.S. Army during World War II (in Italy) and also the Korean conflict. After Korea, he joined the Army Ordinance Corps in which he attended the M-33 radar and Hawk missile schools, was an instructor in missile systems at the Army Missile Command, Redstone Arsenal, Alabama, and was a member of the test launch site for Hawk-Redeye missiles at the White Sands missile range. As an ordinance expert, he was assigned to France to work with the Dorand Company and the Nord Aviation Company in Paris and did field testing and training with the French army on the SS-10-11 missile system. Mr. Criswell served as an advisor to the Korean army during the Vietnam war and for two years as an advisor to the late Shah of Iran. He ended his military career at Fort Bliss, Texas, as the senior operating person of the foreign army direct support unit (Hawk missile firing).

Since retiring from the Army Ordinance Corps in 1970, Mr. Criswell has performed in a variety of civilian maintenance capacities: industrial electrician in factories; general equipment mechanic for facilities and equipment, including steam/chilled water H.V.A.C., at Yale University; maintenance foreman for an electrical generator manufacturer; general maintenance superintendent for a continuous-operation paper, roofing, and siding mill; and human resources manager for several colleges in New York. In addition, he traveled for three years as a service engineer for an (OEM) original equipment manufacturer of machinery for the rubber industry.

Mr. Criswell installed a (PM) preventive maintenance program with great success at Yale-New Haven Medical Center Hospital. He is also a certified energy auditor in the State of Connecticut and participates frequently in energy and maintenance management seminars. He is a member of the Executive Board of the International Maintenance Institute (IMI), Southwestern Connecticut chapter. Mr. Criswell has worked on a number of computerized maintenance and building management systems, both in-house and contractor systems.

Introduction

The only practical route to maintenance that will result in reliable, cost-effective equipment, in facility operation that translates into improved production, in facility-use availability, and in energy conservation is through a planned maintenance program. As anyone knows who has tried to start a realistic planned maintenance program or even to expand an existing one, such endeavors require not only know-how but also long months of both study and trial and error. The pitfalls are numerous for the inexperienced. However, inexperience seldom stops very many from trying, and statistics show that most never succeed. An effective planned maintenance program must rest on three fundamentals: (1) some knowledge of basic operating techniques, (2) manpower management, and (3) effective procedures for turning ideas into workable solutions. These basic elements are embodied into the TEN-step program of this book.

PREVENTIVE MAINTENANCE

The objective of preventive maintenance is to stop functional failure. This objective can be accomplished only by (1) maintaining proven operations (or establishing correct ones), (2) replacing the necessary parts, and (3) lubricating regularly. A faulty operation, a failing part, or lack of lubrication can cause a failure that will, in turn, cause unscheduled loss of equipment, production, or use of an area and necessitate major repair, usually including massive overtime costs. A simple pipe leak left undetected for a short period can cause flooding, ceiling loss, or electrical fires. A slight gas leak left undetected can cause a major explosion. Anyone can cite additional instances of disasters that might have been prevented if someone had just been aware of the problem.

Those who emphasize the use of computers in maintenance programs overlook several elements of an effective program. The computer *can* report the interval of expendable change and the time to lubricate, and it can maintain record files. However, the computer sensor does not see a slight leak; a loose part; a broken, loose, or frayed belt or chain; a backed-up drain; electrical faults; sparking, smoldering, loose connections; burnt-out light sensors; or any sign of contamination. These are only a few reasons for relying on a manual method of preventive maintenance. The computer will report that a failure *has* occurred. However, travel to and from equipment that is scattered or located in isolated areas affords that alert maintenance person a chance to inspect a large area of the premises. Preventive maintenance rounds, properly laid out and conducted by responsible personnel, can be a small but critically important portion of the maintenance day.

Whatever the physical setting—whether in a factory where all equipment is exposed and accessible or in a hospital or hotel where ceilings, walls, and partitions conceal much of the equipment—the key to success in preventive maintenance is coordination. Such coordination must include not only management and supervision of equipment, personnel, and logs and records, but must also encompass the needs of administrators, other workers, and those served by the facility (patients, students, etc.).

An effective, successful planned maintenance program further includes three invariable components: (1) administration approval and support, (2) an adequate budget, and (3) a systematic method for implementing the program.

1. *Administration Approval.* The importance of any program must be firmly established in the minds of those who control the funds for operating the facility. With energy conservation and cost containment foremost in the minds of administrators and private citizens alike, a clear, logical preventive maintenance program becomes even more critical than it might once have been. A proven program like this one—already in successful operation at one site—can provide your maintenance staff with the procedures and know-how to save both money and energy—savings that are likely to draw enthusiastic endorsement from administrators.

2. *A Budget.* Maintenance budgets have an unfortunate history of being traded off for high priority items. Today, however, with the cost of maintaining facilities escalating at an ever-quickening pace, deferred maintenance is, finally, no longer an acceptable practice. The timely installation of a planned, energy-efficient maintenance program, complemented by a professional energy audit, necessarily requires a budget. The first year's budget can be estimated; and as the program progresses, a budget analysis must be a prime objective in taking a professional approach toward energy management and maintenance operations.

3. *The Method: A Manual Maintenance Program.* First, let me explain the difference between a maintenance program and a building management program. The purpose of the management unit is to insure that the facility is in the proper condition for its use or that the equipment is functioning and in the proper state. These tasks are extremely well-suited to a computer operation, especially in a large facility. For instance, the starting or stopping of motors for certain functions, the turning on or off of intrusion alarms for security reasons, and the monitoring of all critical areas for temperature are typical of the functions best served by a computer system. One of the outstanding opportunities for this type of use is in the hotel/ motel business, where the room clerk can turn on the H.V.A.C. from the minimum set-back as he checks in an occupant and then return it to set-back when the occupant checks out. This is an example of realistic building management by computer—with significant savings as a result.

In a maintenance program, however, the only realistic approach is a manual one, controlled completely in-house. Although computers have their place in information storage and retrieval and in reporting malfunctions and failures after the fact, actual maintenance and monitoring must be manual operations. The computer can report a failure once it has occurred; but a maintenance worker on daily rounds—who hears the steam hissing, sees the clogged drain, or notices the bad belt—prevents not only a mechanical failure but often costly downtime, more extensive damage to equipment, or even personal injury as well.

In addition, the manual system is actually the first practical step to a computer system. All the data to be fed into the computer must be researched and written up manually by knowledgeable maintenance personnel. Further, if, after a period of successful manual operation, a facility finds that finances and staffing will allow for a computer to perform records-keeping and clerical functions, the computer operation can then be installed with an experienced staff already in place.

AN OVERVIEW OF THE PROGRAM

This ten-step program (shown as ten chapters) is designed to move a facility systematically from its present level of maintenance and energy efficiency to a higher level that will result in greater profitability through more efficient use of energy and through greater facility and equipment reliability. STEP ONE (Chapter 1) is the program's initiation by management, i.e. administrators. STEP TWO (Chapter 2) defines *maintenance* and assigns responsibility for the program's operation. STEP THREE (Chapter 3) includes a thorough inventory of the facility, equipment, maintenance personnel, and support items to serve as the data bank for all maintenance planning. STEP FOUR (Chapter 4) lists all special-attention items (e.g., fire, intrusion, radiation, contamination, automatically controlled items, and seasonally operated items). STEP FIVE (Chapter 5) establishes the operating arm of the program and includes possibilities for reorganizing present personnel into an organization best suited to your particular maintenance needs and environment. STEP SIX (Chapter 6) describes the job of writing meaningful inspections as the next logical step once an inventory, special requirements, and staff organization have been established. This step also includes assigning priorities to *the work orders* and making them out for each item on the list. STEP SEVEN (Chapter 7) addresses the problems of scheduling on a fifty-two-week maintenance year and of assigning all items on the maintenance list by balancing the workload against available manpower, equipment, and facility requirements. Of course, the availability of materials, equipment, testing devices, etc., for each job is considered. STEP EIGHT (Chapter 8) explores some basic requirements of the maintenance area and possibilities for expansion. STEP

NINE (Chapter 9) puts all the information and activities into usable order through reports and record-keeping, including a logical filing system that facilitates accessibility for planning and reference. STEP TEN (Chapter 10) establishes methods for recording the use and cost of energy and for monitoring energy use and conservation on a continuing basis.

The program lays all the necessary groundwork for a successful planned maintenance and energy-conservation operation. However, no program alone can insure success. To it you must add your own enthusiasm and commitment, your own grasp of the needs of your particular circumstances, and your own high standards for yourself and your staff.

1
Step One: Announcing the Program

Although it must be carried out with the full knowledge and concurrence of the physical plant manager, Step One is really an upper-level management responsibility. In consultation with the physical plant manager, management must decide whether it will implement the entire program or whether only selected parts of it are applicable to its particular operation. Once having made this decision, management's responsibility is to announce to the entire organization that the program will be in place and operating ten weeks after the date of organizational meeting (Step Two).

Announcing a specific date provides a sense of reality and immediacy and gives everyone an objective to work toward. In announcing the program, it is important to enlist enthusiasm and support at all levels of the organization and to overcome whatever reluctance exists to abandoning the status quo. The exact nature of the announcement process may vary somewhat according to the size and resources of each organization. However, the following are some techniques to consider:

1. A meeting with all supervisory personnel, who are then, in turn, to announce the program to their staffs.
2. A description or feature story in the company newsletter.
3. Bulletin board announcements or posters.
4. Announcements on the public address system.
5. Reminder slips in pay envelopes.

A successful program might include a range of these techniques at various intervals in the ten-week preparation period:

Day 1	Management meeting with all supervisors
Days 2-5	Supervisors meet with staffs
Week 2	Announce slogan and/or logo contest for the new program
Weeks 3 & 4	Reminders of contest (and program) using supervisors, house organ, PA announcements
Week 5	Winner announced and awarded small prize
Week 6	Story (with photo, if possible) in house organ featuring prize winner and more details of program
Week 7	Posters, buttons, etc. bearing slogan and/or logo begin to appear
Weeks 8 & 9	Slip with slogan and logo appears in each pay envelope

The importance of generating enthusiastic support among all personnel cannot be overemphasized. All staff must become aware that not only will the program save the company or institution energy and money (then available for other purposes) but also that it will enhance their own productivity as well because of increased efficiency and reliability of vital services, equipment, and facilities.

A final critical consideration for any administrator or manager planning to initiate this program, is the role of the physical plant manager, without whom the program cannot succeed. The common modern business practice of relying on a trained engineer to perform this function has a number of pitfalls. Unquestionably some engineering expertise is essential to the operation of any physical plant, and engineer's training in in-depth investigations is a valuable asset. Few engineers, however, know anything of maintenance people, of maintenance operations, or of maintenance budgeting. A physical plant manager *must* be a manager of people, buildings, and machines, which makes a maintenance person with strong supervisory skills the best choice for this position. Before attempting to implement this program, administrators must have a physical plant manager who has strong maintenance experience, with whom they can communicate comfortably, and who will approach both the program and the maintenance staff with enthusiasm and dedication.

2
Step Two:
Terms and Responsibilities

This step is best accomplished at a meeting called by the physical plant manager. At this time he should (1) define maintenance as work in one of the four categories listed below, (2) run through the steps of the program and the energy conservation list, and (3) assign responsibilities for the program to various leaders and repair personnel. Also at this time, the schedule for the remaining eight steps of the program should be tentatively set: Step Three should begin immediately and should not exceed one month, and Step Four should be done concurrently with Step Three. Steps Eight and Nine should also begin at this time to allow adequate time for studying the organization and for setting up the area and the filing system to be ready at the completion of Step Seven. Step Five should begin at the completion of Steps Three and Four and should be completed in one month; at that time Step Six should begin. All steps should be completed within ten weeks from the date of this meeting.

DEFINING THE TERMS

Both for the overall success of the program and for improved communication in general, it is important for all personnel to understand that maintenance is work that falls into one of these four categories.

1. *Emergency Maintenance* refers to work done on a facility or on equipment that involves personnel safety, material loss or deterioration, or operational delay of equipment. This is unplanned work

9

requiring an immediate response. Although it is impossible to plan for this expense, estimates can be made based on past experience and life-cycle costing.

2. *Preventive Maintenance* is minor work done on a regular schedule to prevent trouble or deterioration to facilities or equipment. This includes such work as monitoring, recording indicator readings or settings, changing filters, draining condensate, minor lubrication, adjustments, and the like.

3. *Regular or Routine Maintenance* includes two categories. Daily services are items called in from or posted in departments and may be either preventive or emergency in nature. In order for any facility to operate, someone must attend to a stuck door, change a burnt out light, move a desk, etc.

Anything of a larger nature is written up on a maintenance work order and planned for. Normally, work that restores a facility or equipment to its original usable condition for a particular application is scheduled and done on a regular rotating basis, whether weekly, monthly, quarterly, or annually. These items make up the other routine maintenance category for this program; that is, scheduled maintenance with necessary parts, tools, materials, and manpower of proper skills and numbers either on hand or under contract. A normal budget item for routine maintenance must be clearly defined, and any exception must be charged to the originating department.

4. *New Work and Projects* include installation, removal, and alteration to existing facilities or equipment. These are normal budget items with approved funding for completion either in-house or by private contract.

ASSIGNING RESPONSIBILITIES

At this time, the responsibilities to be assigned are of two types. First, short-term personnel should be appointed to carry out the functions of installing this program. The data gathering process requires (a) a maintenance person who knows the equipment location and area, (b) a maintenance helper, and (c) a clerk to record the data as it is gathered. Also for the short term are needed (d) a maintenance expert with hands-on experience and (e) a clerk-secretary, both required for inspection writing.

Before the program begins to operate, the following on-going personnel assignments will also need to be made;

1. *The Physical Plant Manager or Plant Engineer* has the over-all responsibility of approving, implementing, and supervising the program. He must insure that the program—whether it is to be adopted totally or only in part—is inserted into the company's procedures with commitment and personal involvement so that no one will attempt any shortcuts. Through consultation with and close observation of supervisors, this manager must keep the momentum that is necessary to install the program and keep it running successfully.

2. *The Maintenance Superintendent or Maintenance Coordinator* must insure that each item of the facilities and equipment assigned to the program is scheduled for required maintenance coverage, including inspections, repair, and records maintenance. In addition, he will inspect each facility or equipment failure to determine its cause.

3. *The Maintenance Repair Department* performs all required maintenance on any item of equipment or facility assigned to this program, in addition to maintaining the complete facility and the equipment of this institution or company. The department must provide a work area, personnel, tools and equipment, and supplies (materials and parts). This is the department that receives planned maintenance orders, completes the work, fills out the necessary action forms, and returns them to the coordinator. In addition, it maintains all reference publications, drawings, and supplies. When items on the planned maintenance list are out of service for maintenance, this department must insure that the next scheduled maintenance (other than daily) is accomplished and recorded.

4. *All Planned Maintenance Personnel* are responsible for inspection, repair, and recording. Preventive maintenance personnel are assigned to rounds maintenance but do not perform major rebuilding or removal of equipment or facilities. They do, however, inspect, test, and adjust; replace expendables; lubricate; clean; spot paint; and record all information. On assigned rounds, this staff must maintain spaces, lights, trash, storage, and security. If repair is indicated, inspectors will determine whether it is emergency or routine in nature and either immediately take the necessary action (if emergency) or complete an action form and report it to the maintenance supervisor.

3
Step Three:
Location, Identification, Inventory

The purposes of this step are (1) to determine what is on hand, (2) to number it according to some usable system, and (3) to record it on paper for ready reference. Sample forms to be used in completing this step may be found in Appendix A on pages 57 through 77.

1. *The Inventory* should include a listing of all of the following elements:

    (a) Human resources (Use Forms 1 and 1A, pages 57 and 58.

    (b) Buildings and grounds (Use Form 2, page 59.

    (c) Equipment and vehicles (Use Form 3, page 60.

    (d) All support drawings and publications (Use Forms 4 and 5, pages 61 and 62.

2. *The Item Location Numbering System* is based on a ten-digit figure, with numbers assigned as described below in Figure 1. Note that this number and description are given by way of example and that specific assignments will vary depending on individual requirements. Numbering of the sites (digits 1 and 2) should begin at some logical point—the main entrance to the facility, for example—and proceed in one direction so that the numbering will accurately reflect the location of the building, lot, etc.

The Item Location Number in the figure reveals that the item is a kitchen in Room 28 on the third floor of the first classroom building. The basic ten-digit format may be adapted to accommodate the specific number of buildings and rooms involved.

A second phase of this location/identification step is to assign numbers 1 through whatever number is needed to all drawings and publications on Forms 4, page 61 and 5, page 62.

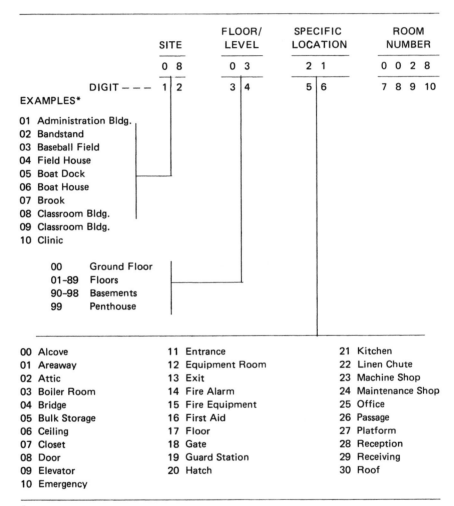

	SITE	FLOOR/ LEVEL	SPECIFIC LOCATION	ROOM NUMBER
	0 8	0 3	2 1	0 0 2 8
DIGIT – – –	1 | 2	3 | 4	5 | 6	7 8 9 10

EXAMPLES*

01 Administration Bldg.
02 Bandstand
03 Baseball Field
04 Field House
05 Boat Dock
06 Boat House
07 Brook
08 Classroom Bldg.
09 Classroom Bldg.
10 Clinic

00   Ground Floor
01–89   Floors
90–98   Basements
99   Penthouse

00 Alcove	11 Entrance	21 Kitchen
01 Areaway	12 Equipment Room	22 Linen Chute
02 Attic	13 Exit	23 Machine Shop
03 Boiler Room	14 Fire Alarm	24 Maintenance Shop
04 Bridge	15 Fire Equipment	25 Office
05 Bulk Storage	16 First Aid	26 Passage
06 Ceiling	17 Floor	27 Platform
07 Closet	18 Gate	28 Reception
08 Door	19 Guard Station	29 Receiving
09 Elevator	20 Hatch	30 Roof
10 Emergency		

*Note that these are examples only and will necessarily be changed to reflect the specifics of any given installation. A manufacturing plant, for instance, might proceed along these lines: 01 Security Gate; 02 Parking Lot; 03 Administration; 04 Laboratory; 05 Change House; 06 Maintenance Shop; 07 Building A; 08 Office Building, etc.

Figure 1. Item Numbering System

3. *The Recording Process* can begin as soon as all items have been inventoried and located. The physical location of each item should be recorded on Form 6, page 63. This record should include a description of the item, its location, and the disconnect/main shut-off location. Each item's location drawing should be entered on Form 7, page 64 or 8, page 65 with a complete entry for its description and location (room or area) and company tag number, if any, included at the bottom of the form in the space provided.

Using Forms 9 through 13 on pages 66 through 73, collect the necessary data and fill in the description and location from the written location forms already filled out. Then proceed to each item and collect nameplate data, together with as much other information on the forms as possible. (The manufacturer, supplier, date of installation, cost, etc., should be available in the maintenance or purchasing files).

Using the identification numbering system described in Figure 2, on the next page, assign numbers to each item and record the number in the following places: data card, written location form, location drawings, and identification number assignment form (Form 14, page 74). Then double check to see that all forms are complete and that all relevant numbers and information have been cross-referenced.

	BASE CATEGORY	CATEGORY BY ITEM	SEQUENTIAL NUMBERING
00 Administration	0 0	0 3	0 0 0 1
01 Athletic			
02 Bulk Storage	1 | 2	3 | 4	5 6 7 8
03 Clinic			
04 Culinary			
05 Design			
06 Electrical			
07 Engineering			
08 Field Service			
09 Fire			
10 Grounds			

01 Air Conditioner, refrigerated	16 Counter
02 Air Conditioner, stm/chw	17 Dishwasher
03 Air Conditioner, window unit	18 Disposal
04 Boiler, steam	19 Door, automatic
05 Boiler, hot water	20 Door, manual
06 Cold Box	21 Door, revolving
07 Compactor	22 Dumbwaiter
08 Compressor, air	23 Elevator, service
09 Compressor, vacuum	24 Elevator, passenger
10 Computer	25 Exhaust Fan
11 Conveyor, belt type	26 Filter
12 Conveyor, bucket type	27 Floor
13 Conveyor, roller type	28 Freezer
14 Conveyor, screw type	29 Furniture
15 Cooking Item	30 Grease Trap

The sample number indicates an administration air conditioner, window unit, Number 1.

Figure 2. Item Identification Number System

4
Step Four: Special Attention Items

In addition to the normal items treated in Step Three, a thorough inventory should be made (on a form like the one at the end of this chapter) of items in the following categories:

1. Items or areas biologically dangerous to humans. The items and dangers should be thoroughly described, as should emergency procedures in the event of exposure.

2. Items that are mechanically dangerous to humans. This includes items not normally operated or not properly guarded as well as items that have an accident history.

3. All fire equipment and its location, including the following: devices, pumps, standpipes, extinguishers, sprinklers, alarms, and monitors.

4. All items that are automatically controlled. For each of these, describe the item, its function, location, controller type, and proper operation. In addition, include the location of the main disconnect (in case of controller failure) and the appropriate procedures to be followed in the event of such failure. Also mention any danger that could occur as a result.

5. All items that must be turned on seasonally, including steam tracers on pipes, steam to air conditioning in summer, and snow melting equipment. Special notes should be made of items that must be adjusted for time changes (daylight savings time) or for natural changes in length of daylight hours.

6. Emergency power equipment, including location, description, sequence of operation, and emergency procedures.

a Planned Maintenance Program

INVENTORY _____
ITEM _____

_____ SIGNED _____ DATE

ONE STEP AT A TIME

5
Step Five:
The Maintenance Organization

The maintenance organization is that positive part of any organization whose purpose is to keep the facilities and equipment in a ready-to-work condition and thereby to insure continuity of operation and/or production. Any maintenance organization must be systematically structured and suited to its own particular work environment. Today, a maintenance organization must, in addition, be energy conservation conscious.

The basic human resources of the maintenance organization are shown in Figure 3. Although any maintenance organization must provide coverage for all of the listed skills, it is common practice in many organizations to use multi-skilled personnel and/or to contract out certain portions of the maintenance work. High-skill jobs—such as computer, electronic and elevator maintenance—and regularly recurring services—such as housekeeping, groundskeeping, laundry, trash collection, and security—are typical items for private contracts, as are purchased steam and chilled water, which eliminate the need for power plant operation. However, a minimum requirement for any organization is that all preventive maintenance and daily maintenance servicing be done in-house. In addition, in-house staff must administer and supervise any and all contractor services.

MAINTENANCE LEADERSHIP

The maintenance leader is a different person at different levels, each with its own priorities and goals.

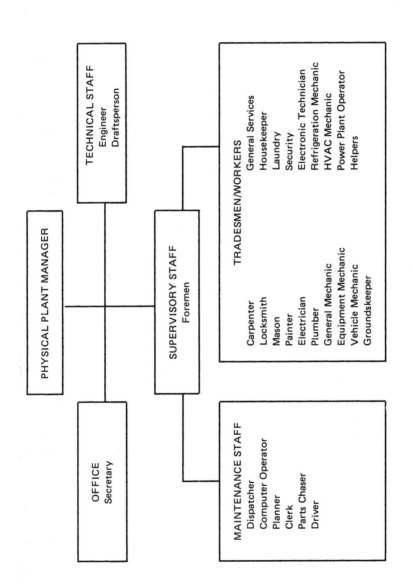

Figure 3. Human Resources of the Maintenance Organization

1. *The Physical Plant Manager* has the overall responsibility of maintaining the facility and equipment in a ready-to-work condition.

2. *The Plant Engineer* advises the administration on equipment and facilities; recommends and inspects all modifications to facilities and equipment; monitors all utilities: electric, gas, water, oil, purchased steam, purchased chilled water, etc. and recommends methods of energy conservation. This leader also monitors the site and drainage, oversees the fire equipment and alarms, and insures that OSHA regulations—as well as city, county, state, and federal rules—on equipment operation are observed.

3. *The Department Engineer* assists the plant engineer in all assignments, research, layouts, new facilities and equipment estimates, modifications, etc. He provides hands-on assistance to the maintenance superintendent for any facilities or equipment problems in his area of responsibility. He also inspects and, in conjunction with the maintenance superintendent, reports to the plant engineer on contract maintenance in his area of responsibility.

4. *The Maintenance Superintendent* is the pivotal member of the maintenance management structure. In this position, the emphasis must be on maintenance procedure and the qualification of skills levels. A maintenance superintendent should have the following background and qualifications: a general education; work experience as a maintenance repairman and foreman; ability to read blueprints and schematics with an understanding of electrical, mechanical, and pneumatic maintenance; a general knowledge of computer maintenance systems and building management computer operations.

The main function of the maintenance superintendent is overall direction of hands-on maintenance; directing foremen; instituting company policy and procedures; setting priorities (with the approval of the plant engineer or maintenance manager) and insuring compliance with them; inspecting the complete maintenance of the company (or branch); and explaining and recommending changes in procedure to the maintenance manager or plant engineer. The maintenance superintendent manages the foremen, work schedules, and plant inspections (equipment, fire, safety, compliance, etc.) and makes all decisions on maintenance matters. He recommends to the physical plant manager or plant engineer any contract maintenance and oversees the fulfillment of such maintenance.

During the course of a typical maintenance day, should a situation arise that is beyond the capabilities of the maintenance personnel, the maintenance superintendent's responsibility is to call in the department engineer and any other technical assistance required to return the equipment or facility to its normal operation.

This leadership position requires hard work, high standards, enthusiasm for and involvement in the job, an understanding of what is going on, and a knowledge of what the problems are. In addition, it requires great tact and a trust and willingness that the maintenance foremen are able to manage their own staffs.

5. *The Maintenance Foreman,* working with his men as a team, is the person who gets the maintenance job done. This position, more than any other, must be staffed by loyal and dedicated individuals, preferably promoted from within the in-house maintenance force. The foreman directs maintenance mechanics in the everyday work on the facilities and equipment in his area of responsibility. He coordinates with other foremen; instructs and assigns repairmen; insures that equipment is available for repair, that parts are procured and available, and that the proper skilled manpower—with the necessary tools—are in place. The foreman should be skilled in at least one craft, with years of hands-on experience with this or a similar maintenance organization. In addition, he must (a) be able to read, understand, and follow various manufacturer's recommendations for maintenance of their equipment; (b) be able to read blueprints and schematics; and (c) know (or learn immediately upon assuming the position) the general operation and maintenance of equipment in his area of responsibility.

Should a foreman receive an order for something that is contrary to his understanding of the needs of the assigned job, he should immediately—and *in private*—bring the situation to the attention of the maintenance superintendent.

6. *The Preventive Maintenance Person* on his rounds is the front line of maintenance manpower. This is the person who, in covering his assigned area, is the eyes and ears of the physical plant manager. He is the one who will see a problem and be able to take effective action. The preventive maintenance person should not perform minor jobs when completing assigned rounds. He should, rather,

go into an equipment room and perform preventive maintenance on all the equipment in that room. Therefore, it is imperative that he understand the normal operation of equipment in order to recognize:

Malfunction	Material Corrosion
Erratic Operation	Leak (air, steam, oil,
Unacceptable Indicator Readings	water, gas)
Material Deterioration	Temperature Variation
Miscellaneous Items	

Further, the preventive maintenance person must have the ability and training to make minor adjustments, do certain tests, change expendables (filters, belts, washers, packings, discs, chemicals, lights), lubricate properly, clean items, and maintain records.

In the course of preventive maintenance rounds, if equipment or facilities are found to need routine maintenance, the preventive maintenance person will write up the work and give it to the foreman, who will assign and schedule the work. The preventive maintenance person can be included in the regular maintenance crew for other maintenance assignments; or he can, upon completion of his daily rounds, move on to other regularly scheduled preventive maintenance work—work that does not demand close daily inspection but that is scheduled for weekly or monthly preventive maintenance.

THE FUNCTIONING OF
THE ORGANIZATION

The smooth operation of a controlled maintenance force is achieved by everyone doing his assigned job. In a functioning unit, the daily operation proceeds along these lines: A *maintenance dispatcher* receives a call that a water pipe has burst. Recognizing this as emergency maintenance, he/she sends all available personnel to respond to the problem and informs the maintenance foreman and manager. The dispatcher then receives a call that a light is out in Room 206, Building Number 3. He/she will then write a daily service job order and post it on the board to assign to the next available daily service mechanic. If this is a critical light (involving essential operation or security), the dispatcher will either send the next available person or request a mechanic from another job (with his foreman's permission of course) to perform the job. A call to have a lighting fixture moved, is written up by the dispatcher on a planned maintenance form and given to the planner/electrical foreman. The foreman will then evaluate the request and—if it is found to be valid —list the material, tools, manpower, and estimated time for the job. In addition, he will list the telephone number and name of the contact person to coordinate such items as area accessibility. The job is then scheduled and placed on the projects board; the individual foremen are notified; and parts are ordered. Upon completion of any job, the foreman responsible must personally inspect the job and accept the results.

Throughout the normal operation, the *maintenance clerk* should maintain records, fill out various reports, receive requests for supplies, complete necessary paper work, assign correct P.O. numbers and project (J.O.) numbers, and follow through on all such items until received and signed for by himself. The maintenance clerk should maintain the equipment data file, the fifty-two week maintenance schedule, and the equipment location and numbering files. He should also issue the planned and preventive maintenance work orders to the dispatcher, who distributes them and posts all planned work on the project board. Rounds work orders are assigned by the appropriate foreman, and the rounds inspection forms should have a hanging position in the dispatch room for storage. The building management computer operator (who could be the dispatcher or clerk) must monitor the computer, respond to its indications (e.g., turning off the boiler or security alarm; alerting security, fire department, and/or maintenance staff), and must also know the facility, the location of the sensors, and what the sensors control or monitor.

COMMUNICATION WITHIN
THE ORGANIZATION

In order for any organization to function effectively, the people involved in its operation must be kept informed of what is going on and must be asked for their help and cooperation on all job-related matters. Management should meet with foremen at least weekly and should encourage them in any and all situations that warrant it. Foremen should hold a brief meeting (maybe as short as five minutes) at least weekly to talk with their staff. Management, foremen, and workers should routinely meet at least once a month; and management, foremen, and workers should hold annual meetings of an hour or so. All major decisions should be announced publicly, and constructive suggestions should be encouraged.

An enthusiastic spirit in any organization is achieved over a period of time through the working together of people for the common good and through recognition for jobs well done. Awards, incentive, and training programs are valuable stimuli toward continued performance; and rewards—both verbal and financial—often spur workers to even greater dedication and productivity.

The importance of effective communication, however, extends beyond the maintenance organization itself and must be expanded to include user/operator awareness. Negligence and misuse by personnel assigned to operate facilities or equipment often result in enormous expenditures of both time and money in repairs and/or replacements. The importance of turning off equipment that is not in use and of operating it at the appropriate rate and under the appropriate conditions must be communcated to personnel throughout the organization.

A Special Note

If you are a one-man work force (e.g., building superintendent) or part of a very small work force, take heart. This program if you follow through all the steps, can still result in better productivity and reduced costs. It may take you longer to complete the inventory phase (or you may be able to afford a few students for part-time help in this phase), but the system will yield both energy savings and better overall results in your work.

6
Step Six:
Inspections and Job Orders

This is the point at which the information gathered in Step Three is entered on the necessary rounds inspection and job order forms for each item of facilities and equipment, beginning with Category 01 and moving sequentially through the list, making sure to complete all requested information on each form. All forms, some inspections, checklists, and guidelines for planned maintenance are found in Appendix B (pages 83 through 123).

Inspection rounds provide a systematic look at the items on the maintenance list and at the facility area as well. If rounds assignments are carefully laid out, all areas of the facility will be seen by the rounds personnel over a given period of time. It is on these maintenance rounds that minor lubrications, filter changes, minor adjustments, condensate draining, indicators inspections (gauges, sight glasses, lights, etc.), and automatic control device checks are performed.

The rounds form (Form 18, page 83) maintains a schedule for the necessary jobs. Any work done during rounds that exceeds the set time for the rounds must be written up on a maintenance action form (Form 19, page 84), which is also used for reporting an equipment failure. This form shows the time and material used beyond the normal rounds requirements. (Of course the rounds person can see problems before they become failures.)

Energy conservation and preventive maintenance rounds must include all of the items listed in Figure 4.

ENERGY CONSERVATION AREA	PREVENTIVE MAINTENANCE ACTION
steam leaks	detection, adjustment, job order
water leaks	detection, adjustment, job order
gas leaks	detection, adjustment, job order
air line (compressed air) leaks	detection, job order
unnecessary light	detection, turn off
insufficient light	detection, fixture cleaning, bulb replacement
equipment, motor, light, automatic control exceeding cut-off time	detection, adjustment, job order
equipment, motors, etc., running unnecessarily	detection, turn off
HVAC not heating, cooling, exhausting, etc.	detection, adjustment, job order
excessive machinery use of lubricant, coolant, drive power; abnormal or erratic running of machinery	detection, adjustment, report
steam trap leaks	detection, job order
insulation loss	detection, addition, replacement, job order
access doors, windows, screens, vents incorrectly positioned, loose, or damaged	detection, adjustment, job order
roof leaks, clogged drains, gutters, loose or missing flashings, etc.	detection, adjustment, job order

Figure 4. Energy Conservation on PM Rounds

Tour by preventive maintenance personnel should comprise not only equipment but also the entire facility—including roof tops, stairwells, connecting bridges, walkways, areaways, fire escapes, parking lots, roadways, fences, swimming pools, storage areas, outside lighting, loading docks, and fire and alarm systems and devices. (See Inspections 1, 2 and 3.) Rounds should, therefore, be laid out to insure area coverage as well as item coverage. For instance, in a building with five stairwells that is connected to the main complex by two

PLANNED MAINTENANCE	MAJOR ITEM	PREVENTIVE MAINTENANCE
Install New Cover Seal Flashing/Cornice Install New Gutters Point Chimney/Brickwork Inspect all Equipment Mounts and Vent Pipe Openings	ROOF	Spot Repairs Clean Inspect/Clear Drains & Gutters Inspect for Snow/Rain/Ice Inspect for Bird Nestings
Paint Point Sandblast Resurface	WALL EXTERIOR	Spot Paint Minor Repair Clean Inspect for Deterioration
Replace Paint Resurface	WALLS, FLOORS, CEILING INTERIOR	Spot Paint Clean Inspect for Deterioration
Replace Wash Store Install Paint	WINDOWS/SCREENS/AWNINGS	Inspect Replace Panes Replace Hardware Spot Paint
Annual Clean Out Septic Tank Annual Storm Drain Clean Out Floor Drains/Chemical Treatment Slop Sinks/Chemical Treatment Laundry Drains/Sumps/Chemical Treatment Drain Water Tanks and Clean Adjust/Operate all Major Valves/Repack as needed	PLUMBING	Chemical Treatment (Minor) Clean Strainers/Filters Replace Washers Replace Packing Adjust Flow Controls Insure Expansion Tanks/Pressure Level, etc. Correct
Clean/Rebuild/Adjust Calibrate Controls Test Operate Entire Unit including all Controls	HEATING/COOLING	Clean Filters/Strainers Adjust Controls Test for Chemical Additives Inspect all Devices/Piping Minor Repair Verify Fuel Supply Read Indicators/Gauges/Sightglass Verify Controlled Space Temperature

Figure 5. Facility Maintenance Items

PLANNED MAINTENANCE	MAJOR ITEMS	PREVENTIVE MAINTENANCE
Planting Trimming Pruning If Fruit, Pick Rake Leaves	TREE/BUSH	Water Cultivate (root area) Spray Chemical Treatment
Plant Pick Cover Remove	FLOWERS/BUDS/ OTHER	Water Weed Spray/Dust
Plant Fertilize Cut/Trim	GRASS	Water Weed/Spray Spot Paint Sweep/Clean
Install Curb Resurface Oil Paint Remove	ROADS/SIDEWALKS/ PARKING LOTS/ SPORTS AREAS/ OTHER CEMENT/ BLACKTOP AREAS	Spot Repair Spot Paint Sweep/Clean Align Curbing Repair Bumpers
Install Fabricate Paint	SIGNS	Minor Repairs Clean Align Tighten/Adjust
Install Remove Modify	LIGHTS	Inspect Fixture/ Mounting/Tubing/ Wiring Clean Align Change Repair Minor Damage
Install Major Repair Paint Remove	FENCES/WALLS/ GATES/DOORS	Minor Repair Straighten Spot Paint Clean
Clean/Paint/Test	DRAINS/POOLS/ FIRE HYDRANTS	Emergency Clean/ Spot Paint

Figure 6. Grounds Maintenance Items

walkways, the rounds personnel should enter the unit by way of one connector and return by the other. While at the unit, the preventive maintenance person should go up a different stairwell each day of the week. This routing provides good coverage of little used but vitally important areas.

The manhours required for thorough preventive maintenance rounds must take into account the following factors:

1. *Preparation Time* includes the time required for instructions and work orders; for obtaining lubricants, tools, supplies, reference materials, blueprints, and permits; and for notifying the user.

2. *Travel Time* must allow not only for distance but also for mode of transportation and whether or not elevators, vertical ladders, stairways, or tunnel are involved.

3. *Restrictions* may involve transportation availability, a locked area, or occupant use of the facility or equipment.

To these variables must be added the following considerations:

4.   Actual inspection of equipment
5.   Performance of necessary maintenance
6.   Clean-up

The inspection forms, job order forms, project orders, and maintenance action forms record the total manhours, materials, and associated cost of the actual work done by the maintenance force. It is extremely important to understand the extent to which the total effort depends on maintaining these forms and reporting all work. All these items are necessary to determine budgeting for the department, as well as to assign responsibility and to modify building and grounds drawings and equipment operation procedures.

The rounds form can be used on monthly rounds for filter changes, etc. if you have a sufficient number of items to justify these rounds.

The maintenance work order (Form 20, Appendix B) is to be filled out daily on call-ins. It is done in two part, the original is kept on file and the carbon copy is given to the repair person.

Work Request/Job Order/Unplanned (Form 21, Appendix B) is to be filled out for any work found by inspection or failure.

Planned Maintenance Job Order (Form 22, Appendix B) is to be filled out for all planned maintenance other than projects.

Project Order (Form 23, Appendix B) is to be filled out on all new installations, modifications and removal jobs.

Form 24, page 101, and Forms 25 through 29 on pages 119 through 123 are for your information.

7
Step Seven: Scheduling

Scheduling is the means of bringing together all the previous work by taking all known information and causing work to be done in an orderly manner to assure that all items receive the maintenance necessary for continued reliable operation. Effective scheduling must consider the following factors:

the jobs to be done
the skills (craftsmen) required
the skills (craftsmen) available
the manhours required (by craft)
the manhours available (by craft)
the availability (hours, day, etc.) of items requiring work
the spare/repair parts available
the operators, instructions, drawings, tools, and the testing materials available
the importance of the item to normal operation of the facility
the availability of emergency support system(s)

Once these determinations have been made, scheduling can begin, using Form 30, Appendix C, page 127. In the left-hand column of the chart begin with the first category and first number and proceed listing in sequence until all numbers are recorded. Then, to the right, lay out fifty-two squares to represent the fifty-two-week schedule this system will follow. From this schedule, we will assign all rounds (other than daily rounds) and job orders, and list each on the master schedule in frequency spots (week, month, quarter, etc.).

Once all items are listed, the work is prorated to balance the workload. (Keep in mind any special requirements, such as seasonal time.)

With the basic system established, a meeting of the supervisors directly responsible for maintenance of the various items will review the inspections and job orders. After an explanation of the concepts, the work required, and what is possible in the given circumstances, supervisors should receive individual copies to study and critique for accuracy and feasibility and for recommendations for improvements. Private follow-up conferences with individual supervisors provide an opportunity to concentrate on specific areas. After a second group meeting (perhaps a week after the first) at which critiques are completed and decisions on the workload are finally modified, the master schedule should be revised to reflect the loading frequencies.

A fifty-two week file should contain a separate folder for each week, each with an index page showing what work is in the folder, the section assigned to complete it, and the frequency (see Form 31, page 128). The priority of maintenance, identification numbers, assignment numbers, locations, types of pieces, and all other relevant information should be entered on Form 32 (page 129)—the register for frequency of maintenance. Further scheduling required to operate the system is the scheduling of craftsmen to the jobs on a form similar to the fifty-two week one (see Form 33, pages 130–131). The jobs to be done in the scheduled week are listed by category and sequence number at the left. Then all craftsmen (carpenters, mechanics, etc.) are listed across the top. The space below each craft is divided into available manhours and assigned. Proceeding down the page, the number of available manhours is reduced by the number already assigned. This comprehensive scheduling of maintenance work—done by the dispatcher/scheduler in collaboration with the foremen—gives the foremen a chance to defer and catch up on work as the maintenance workload varies.

No job is complete until the final inspection is completed and the equipment or facility is functional and secured (guards in place, insulation replaced, spot painting done, signs replaced, old parts removed, and area cleaned). Should additional deferred maintenance be required, it should be properly recorded and reported. (See Form 34, page 132).

8
Step Eight:
The Maintenance Area

That portion of the physical plant occupied by maintenance must include both office space and a shop area that will allow for making up equipment and for dismantling pumps, motors, compressors, air conditioners, and all vehicles. An area should be devoted to each of these categories: HVAC; electrical; electronic; painting; carpentry; and records, drawings (blueprints), and reference publications. The maintenance area must include desk space for individual foremen in their area of work, locker room and toilet facilities, and a storage area for the various items associated with the job.

Storage frequently demands special conditions and precautions. Paint and wood require controlled storage, usually in bulk; some small amount of metal should be racked; the electrical, plumbing, HVAC, electronic, locksmith, and door and window hardware, as well as general supplies should all have designated areas; and, of course, the tool room must insure that control and storage are good.

One special consideration for most maintenance departments is storage at distant locations from the maintenance shop. It is sometimes wise to have at the location a ladder, an industrial (wet-dry) vacuum, and certain other supplies that are used there on a regular basis (filters for air conditioners, light bulbs, etc.). These satellite storage areas allow the preventive maintenance person to move about the facility without a large cart with supplies and a ladder. Some places use work vehicles (trucks or vans) equipped with such items. The staff that performs regular maintenance jobs will have their carts and other supplies with them, but the front line of the maintenance force—the preventive maintenance rounds person—must be able to move about the facility as unencumbered as possible and benefits greatly from satellite storage areas.

SPECIAL AREAS

The machine makeup repair department should have all the various electrical services it needs (110, 220, 440, single phase, and three phase) with proper controls and means of hookup. It should also have an air supply (blow-off, inflating, and device operating), a degreasing tank, welding area, truck loading area or dock, and lifting devices. In addition, it should stock such shop tools as a jigsaw, grinding wheel, metal cutoff saw, drill press, power press, high pressure washer (portable for site use when needed), and spray.

The carpenter shop should have a table saw, rip saw, planer, drill press, power sander, jigsaw, panel cutter, and a good exhaust system. This is one area that demands space—both for working with large pieces of wood and for storage of supplies—and it should have access to the outside.

The paint shop is an area that does not require a large space but does demand isolation, a storage area, and, of course, proper ventilation. It must include a slop sink and water supply.

The central area of the maintenance operation is occupied by the dispatcher, who must be somewhat isolated yet still accessible through some partial barrier (window, half door, or the like). In his/her control room is the maintenance action board, and on it is posted all the work presently in progress. Here too is the project board with all the planned work on it; a master listing by job title of all personnel assigned to the department; emergency phone numbers (and beeper numbers, if applicable). The vacation schedule and daily absentee list should also be posted in this room, as should all emergency procedures. The building-management computer, if any, is located in this control room, which is also the storage center for keys and off-duty beepers.

A special equipment locker should house such items as Scott air packs, fire tools, and foul-weather gear. Adequate space must also be provided for the storage of machines used by the maintenance department (trucks, snowplows, portable compressors, etc.). All waste generated by the maintenance force must have adequate storage and disposal areas.

SUPPLIES

Any maintenance organization must store basic supplies. The decision of what items to stock should be based on item availability, past history of use, and—in the case of major items—life expectancy. All supplies should have a secured area and should be separated by category and properly stored in shelves, bins, skids, drawers, etc. All should be properly labeled and recorded in a current inventory system. Items that are to be stored outside should be properly protected from weather and theft; items stored in dusty areas should have dust covers. All gears should be lubricated and taped over; all threaded areas should be taped over for protection; and all items requiring it should receive a grease coating. Finally, all openings on stored items should be closed or covered to keep out dirt and foreign objects.

All tool room items should be properly stored and inventoried. Chemicals should have special handling and storage, at least to the extent of OSHA requirements and the dictates of common sense.

A special area of interest in today's equipment is damage by static electricity to electronic components and boards, largely due to negligent or uninformed handling by the store room personnel or by a repair person. If precautions are not taken to prevent such mishandling, extremely costly damage may result.

Another critical area of supply maintenance is a thorough, up-to-date record of the manufacturers and parts suppliers for equipment. Accurate records of this type frequently prevent costly delays in maintenance or in replacement of equipment. The average cost of a purchase order is staggering and is brought about in large part by trying to locate the vendor and/or the part at the time of need (breakdown).

9
Step Nine:
Reports and Records

When the work is done and job orders are properly filled out with all required information, they are returned to the maintenance dispatcher, who routes them to the maintenance clerk, who then extracts the various items of information and files them for future reference. This planned maintenance program uses a *monthly action report* (Form 35, page 135, Appendix D) to extract the volume of work performed, the material and manhours used, the types of inspections performed; the costs of materials, and the exact items worked on during the month of the report. The *file* is set up by frequency of inspections (weekly, monthly, quarterly, etc.); and a section is noted as "Weekly Inspections Completed," "Monthly Inspections Completed," etc. These files are further separated by maintenance section (i.e., mechanical, HVAC, electrical, etc.) so that all electrical work is in its own section of the frequency file. For instance, all maintenance done on a quarterly frequency by the electrical department is in the quarterly file under "Electrical."

The next file to be set up is the *action form file.* This file is organized by building, starting with building 01, with a thirty-one day separation to file any action forms completed on the equipment or facility. These are filed on the day the work is *initiated,* regardless of the month. Another file logs *rounds forms returned* by frequency and by section assigned. Another file is for all *contractor work* and should be organized by contractor and by the frequency specified in the contract.

The necessary files for the maintenance department should include—but are not necessarily limited to—the following:

Personnel
Job Descriptions
Vacation Schedule
Accident Reports
School Information
Contractor Maintenance Contracts
Vendor Addresses, Correspondence
Data Bank
Master Schedule
Completed Job Orders
Action Forms
Equipment Failures
Monthly Action Reports
Maintenance Reference

Maintenance Drawings
Planned Maintenance
Safety (Accident or Incident
   Reports)
Vehicle Operators
Purchased Gas (Acetylene,
   Oxygen, etc.)
Vehicle Oil, Gas, Diesel, etc.,
   and Batteries (commercial)
Items Delivered on Regular
   Basis
Special Attention Items
New Work
Requests for Estimates

ADDING AN ITEM TO THE PLANNED MAINTENANCE PROGRAM

To initiate this action, a planned maintenance request must be submitted to the planned maintenance coordinator, using Form 36, Appendix D, page 137. The necessary information-gathering forms must be completed by maintenance personnel and returned to the planned maintenance coordinator. Requests for any necessary publications should be prepared. Upon completion of the above procedures, the scheduling of manpower, supplies, and equipment can begin immediately.

RECORDS PROCESSING

Accurate and effective records require that the following steps be taken:

1. The maintenance person must fill in all assigned work orders, rounds, check sheets, and inspection follow-up forms (Action Needed/Action Taken).

2. The supervisor must assign and *insure* completion of rounds, repair, or follow-up information to the maintenance department for necessary scheduling and must also forward the forms to the planned maintenance coordinator.

3. The repair department must make out job orders, forward the job order number and date to the planned maintenance coordinator on an action form; schedule and complete action, send form to the planned maintenance coordinator.

NOTE: All equipment failures must be reported to the planned maintenance coordinator, together with a report of the work required to repair—including time, parts, and manpower. The Action Needed/ Action Taken form should include the time the equipment went out of service and the time at which service was restored.

Miscellaneous forms to be used by the section and included are Forms 37, 38, 39 and 40.

10
Step Ten:
In-House Energy Management

This series of practical energy conservation recommendations does not involve making major modifications, purchasing new equipment, or otherwise expending large amounts of capital. Certainly a professional energy audit—should you have the funds available—will pinpoint your energy use and recommend further conservation steps to include retrofitting and purchasing new equipment; and it can be a valuable planning tool. However, this section simply lists practical steps toward conservation.

Energy conservation begins with the awareness and personal involvement of all personnel. Without widespread—if not total—cooperation, the measures cannot succeed.

RECORDING USE AND COST

To initiate the conservation effort, record all present uses and costs and compare them against last year's figures. A log set up by month and type of cost—with columns showing last year's figures, this year's figures, and spaces for continuing the record into the next few years—will clearly show what is happening and will encourage comparisons to reveal the results of your efforts.

I. Energy Use
    A. Electrical Consumption
        1. List either KW or KWA by month as billed.
            a. Record total cost this year and last.
            b. Determine demand and/or ratchet clause.

      c. Explain any difference in billing.

      d. Describe the types, numbers, and locations of meters used to measure consumption in the building(s).

      e. Identify the amount of emergency electrical power (generators) that is in-house or on call.

B. Natural Gas Consumption
1. List thousands of cubic feet (MCF) or therms by month.
2. List location of meters.
3. List total cost.

C. Fuel Oil (by month)
1. List the type.
2. List the gallons.
3. If a mixture, list mixture and amount used.
4. List total cost.

D. Coal (by month)
1. List by type.
2. List number of tons.
3. List total cost.

E. Purchased Chilled Water (by month)
1. List amount in ton/hours (or other, if used).
2. List demand charges.
3. List total cost.

F. Purchased Steam (by month)
1. List amount used in thousand of pounds at what psi.
2. List demand charges, if any.
3. List total cost.

G. Vehicle Fuel (by month)
1. List by type.
2. List gallons used.
3. List total cost.

NOTE: Have the local utility check your power factor (PF); this information will allow your efforts to be even more effective if correction is needed. In addition, take (or have a professional take) infrared photographs of your facility during the heating season to provide visible evidence of energy leaks.

II. Energy Conservation
Use the inventory form on page 18 (Chapter 4) to list each item and corrections. Sign and date the form.

A. Boilers
  1. Insure that operators understand correct operating procedures.
  2. Clean boiler.
  3. Adjust burners.
  4. Adjust blowdown.
  5. Adjust pressure for load (summer/winter).
  6. Insure that stack monitor is working and clean.
  7. Insure that water chemical treatment is correct.
  8. Return condensate on steam units to save water, chemicals, and fuel. (Be sure condensate return temperature is acceptable.)
  9. Meter condensate returned (flow) and temperature. (Record by shift.)
  10. Meter fresh water use and temperature; if not preheated, preheat fresh water. (Use either condensate return or heat from blowdown energy saver output.)
  11. Meter fuel use by shift.
  12. Chart chemical use by shift.

B. Steam/Hot Water (Piping System)
  1. Traps
     a. correct operation
     b. properly sized
  2. Valves, gaskets
     a. leak and corrosion free
     b. valves operated to insure correct operation.
  3. Bypass valves should not be blowing by.
  4. Hand holds should not be leaking or show signs of deterioration.

5. Expansion tanks
   a. deterioration
   b. leaks
   c. fill level for fluid and air.
6. Flash tanks should not show signs of deterioration.
7. Condensate receivers (no signs of deterioration).
8. Piping connections, hangers, fitting (no deterioration).
9. Insulation.

C. Domestic Hot Water
   1. Turn temperature down to 110F at end of line.
   2. Meter use.
      NOTE: The instant heater type is preferable to the large tank; an in-house heat plant can provide the heat source, rather than electric or gas.

D. Laundry Water
   1. Turn temperature down to 150F.
   2. Insure that unit is cycling correctly. (Wash, rinse, etc.).
   3. Water levels and rinse flow should be properly adjusted.
      NOTE: Wash water discharge is a heat source; if necessary, install a holding tank to preheat fresh water.

E. Equipment Size and Cycle
   1. Insure that every item is the appropriate size for its job.
   2. Duty cycle all items that its use will allow.
   3. Insure that every operating item is off when not in use or not needed.
   4. Insure that all electronic, electric, mechanical, and pneumatic adjustments are correct.
   5. Insure that all items such as drive belts are the correct size and that matched sets are used where required.
   6. Inventory all filters and strainers; insure that they are in place, clean and functioning correctly.*

*This is a priority preventive maintenance item.

7. Insure that all automatic lubrication systems are in place, functioning correctly, and regularly supervised.*

8. Attend to other items in Step Four of the program.

F. HVAC

1. Use the inventory items in Step Three to locate.

2. Insure that all filters are in place, sized correctly, and clean.*

3. Insure that all heating and cooling coils (including preheats and reheats) are clean and clear of obstructions.

4. Insure that all dampers operate correctly and have a good seal.

5. Insure that all controls are calibrated and functioning correctly.

6. Revise all cooling and heating schedules to lowest heating level and highest cooling level.

7. Shut down unneeded air handlers.

8. Balance air handlers.

9. Minimize heat to stairwells, storerooms, and little used areas.

10. Insure that all chimneys or stacks not in use and without automatic dampers have their openings closed and properly posted.

11. Shut down heating to air conditioners when outside air is above 65F; shut down cooling when outside air is below 65F.

12. At outside air temperatures 50–60F, hook up perimeter hot water heating to chiller cooling system if appropriate.

13. At outside air temperature 50–60F, use outside air for cooling in hot areas of the plant.

14. Any high ceiling areas with stratified air, use fans to drive hot air down; if air is smoky or contaminated, use filtered duct suction system to return air to lower level.

---

*This is a priority preventive maintenance item.

G. Lighting
   1. Inventory.
   2. Lower levels where appropriate.
   3. Clean fixtures, reflectors, lenses, and protective covers.
   4. Use energy-efficient bulbs and ballasts.
   5. Replace incandescents with fluorescents.
   6. Use automatic controls (photocells, timers, computers, and ultrasonic sensors) where appropriate.
   7. Have all employees turn off lights when not needed.
H. Emergency Electrical Power (Generators)
   1. Determine peak periods of use by monitoring service in.
   2. Determine what items can be turned off during peak-use times; if necessary, operate at non-peak periods.
   3. Determine what items must be used during peak-use times and select appropriate items for load of emergency generator; generator use during this period will save the cost of high demand from peak use.
I. Windows, Doors, and Hatchways
   1. Inventory windows by amount of surface area, type, and location.
   2. Inventory all outside doors by type (single, double, revolving, sliding, etc.).
   3. Inventory all hatches by type (automatic, manual, wood, metal, etc.).
   4. Check infiltration and convection loss due to cracks, bad closures, etc. (Use infrared photos if possible.)
   5. Seal windows, doors, and hatches with weatherstripping; insure proper covers on all vents.
   6. Have all window coverings closed at night in the heating season; insist that north-facing windows be closed and insulated if at all possible.
   7. All receiving docks should have dock closure doors and be isolated from main heating areas.

8. Doors without automatic closures should be on an alarm system to indicate when they remain open.
9. Weatherstrip all doors. All except essential doors should be operated from inside only (no handle or keyhole on outside).
10. Enclose pedestrian entrance doors with glass if possible.
11. Use awning shades on outside of windows where possible.
12. Trees or walls on sides of buildings with north or northwesterly exposures or prevailing winds provide additional insulation.

J. Roofs
1. Inventory all roofs.
2. Determine condition of roof, flashing, equipment footings and bases, cornices, drains, gutters, electrical grounding systems, etc.
   a. Point all brick.
   b. Repair or replace flashings.
   c. Repair or replace equipment footings, etc.
3. Use rented roof-spraying equipment to refinish roof surfaces if necessary.

NOTE: Roof drainage is a notorious troublemaker. Drains stopped up with a bird's nest can flood and then freeze in winter, causing separation of the building itself. Stopped-up downspouts have caused water to pool and break hangers on gutters, and ground pools have caused basement flooding—especially destructive in northern regions where the standing water freezes and causes separation.

III. Special-Attention, High-Cost Items (Users and Operators)

A. Insure that each operator/user understands the correct operation of equipment, can recognize deterioration or faulty operation, and knows who to call for assistance.
B. Insure that area lighting is off or at low levels when not needed.

C. Insure that all windows, doors, and hatchways are closed (in any conditioned area) and that they have a good seal.

D. Insure that thermostats and heat controls are not tampered with and are kept at minimum settings.

E. Insure that water faucets and toilets shut off and do not leak or drip.

F. Discourage the costly, unnecessary use of elevators by employees.

ON-GOING MONITORING

Once the measures detailed in the previous section have been taken, your monitoring of energy consumption must continue if the program is to be effective. The number of the following recommendations that are feasible for any particular installation may vary somewhat, but, whatever your situation, some kind of surveillance of energy use and users is essential.

I. If you have the staff and budget to do so, establish a permanent *Energy Control Office* to perform as many of the following functions as possible:

A. Hire a professional energy manager.

B. Use energy survey and recommendations.

C. Use a preventive maintenance program.

D. Distribute the enclosures in Appendix E to assure personnel involvement in the program.

E. Hire a controls technician (HVAC).

F. Use public display boards in lobbies and near cafeterias, locker rooms, etc.

G. Organize energy displays and fairs. (Use willing vendors, utilities, and energy office personnel.)

H. Issue citations for violations and awards for items brought to the attention of the office.

I. Have top administration continually refer to energy measures.

J. Use the bulletin (house organ or newsletter) to keep the community informed of energy measures.

K. Insure that energy monitoring is constant—for all days and for all shifts.

II. HVAC

A. Calibrate all controls and verify total unit operation. NOTE: Check all dampers.

B. Turn off steam (or chilled water) in season.

C. Maintain a filter-change program.

D. Calibrate cooling tower water.

E. Use economizer on air handling. (In the case of return air, insure that thorough filtration—including odor—is accomplished.)

F. Use extreme caution in areas where bacteria could contaminate the air.

G. Insure preventive maintenance for all refrigeration—including, but not limited to, refrigerators, freezers, ice-makers, walk-in boxes, environmental rooms, and refrigerated water coolers.
   1. Blow out compressors.
   2. Defrost and clean evaporator.
   3. Check door gaskets and closing hardware.

III. Plumbing

A. Adjust all water systems for pressure, flow and temperature.

B. Install economizers on all showers, flushometers, and toilet tank units.

C. Check all outside and isolated water connections and sprinklers.

D. Institute a steam trap program.

    E. Turn off steam (or chilled water) in season.

    F. Inspect all roof drains.

    G. Inspect all floor drains.

    H. Thoroughly inspect all vent piping.

    I. Open and close (operate) all valves at least once a year.

    J. Change all water filters (cartridges) regularly.

    K. Calibrate all gas-burning equipment (stoves, ovens, hot water heaters, stills, etc.).

IV. Electrical

    A. Megger all motors of five horsepower and above at least annually.

    B. Inspect, clean, and adjust all d-c motors, including commutator, brushes, coils, and controllers.

    C. Inspect, clean, and adjust all photocells.

    D. Inspect, clean, and adjust all timers with seasonal time changes.

    E. Inspect all fuse holders.
       1. Check for burnt or dirty contacts.
       2. Check for proper fuse type and size.
       3. Check continuity (resistance) of old ones.
       4. Blow out cabinets.
       5. Tighten connections.

    F. Measure lighting levels in all areas and adjust as necessary.
       1. Use energy-efficient bulbs and ballasts.
       2. Set known areas on timers.
       3. Use re-lamp program.
       4. Remove all incandescent bulbs and replace where required with new circleline units.

    NOTE: All lights should be observed by security, housekeeping, grounds personnel, and engineering staff.

    G. Determine elevator exciter units turn off when not in use.

V. Carpentry

   A. Insulate all door and window casings and facings.

   B. Inspect all door closers and install automatic closers in all practical locations.

   C. Install proper drapes, blinds, or other coverings on all building exterior wall glass and establish a program (Housekeeping perhaps) for opening and closing them.

   D. Inspect and repair all exterior shell (building) damage: broken windows, door; roof damage; grounds drainage to the building; downspouts; etc.

   E. Inspect and repair all stairwell openings, doors, windows, and hatchways.

11
Computerizing Your Maintenance Operations

When you have completed the ten chapters of this book that are dedicated to guiding you through the necessary steps to set up a realistic maintenance program, you will be properly prepared to begin planning how to place your maintenance operations on a computer maintenance management program.

You have developed the necessary data base for computerizing your operations and it will be relatively easy to mate your ten-step program to a computer system.

Important items to consider:

1. What computer and associated software will accomplish your objective?

2. Exactly what data will be needed?

3. How is the system operated?

These are must items to be considered before purchasing any hardware or software.

Here in this new chapter you will learn how to computerize your maintenance operations. The number, types and brand names of computers now on the market are endless and choosing the right software is a formidable task.

IMPORTANCE OF TRAINING

The computer is a very sophisticated tool and will require personnel who install, operate and supervise its operation to be properly trained before you can expect the system to be on line and operating in your physical plant. Computer operation requires attention to detail. There is a definite period of learning for you and your staff that is a critical part of computerizing your maintenance operations.

STEPS IN COMPUTERIZING YOUR OPERATIONS

Identify Goals

The first question that you must answer is: What are your goals for the system (what do you want the computer to do for you)?

Remember that the most important aspect of computerizing your operations is to find the right software. The software usually determines which computer you will purchase.

In developing system specifications you must establish your requirements in detail such as:

A.   Maintain your maintenance-dependent item inventory.

B.   Maintain the data bank of information on each item.

C.   Maintain required maintenance analysis file on each item.

D.   Maintain parts inventory.

E.   Maintain materials inventory.

F.   Maintain the maintenance reference library inventory.

G.   Maintain the maintenance location maps file.

H.   Maintain the special tool inventory.

I.   Maintain the test equipment inventory.

J.   Maintain the equipment inventory (backhoes, snow blowers, grass cutters, etc.).

K.   Maintain the vehicle inventory.

L.   Maintain the department vacation schedule.

M.   Maintain the master (52) week planned maintenance schedule.

N.   Maintain the vendor's list.

O.   Maintain the manufacturer's list.

P.   Maintain the contractor's list.

Q.   Maintain employee roster.

R. Print planned maintenance job orders automatically at right time (frequently on demand).
S. Print daily — work request, job order on demand.
T. Print daily activity report.
U. Print monthly activity report.
V. Print monthly financial report.
W. Print purchase order request automatically or on demand.

Collect Computer Information

With this list of goals in hand it is time to assemble manufacturers' information. Call manufacturers of hardware and software and find out what the possibilities are for you to get a system that can accomplish your objectives. When deciding on the program (software), you must be sure the program is compatible with the computer. Also check if the software can be modified to include the special features of your operations. Use inhouse specialists or outside consultants to help you evaluate how your requirements can best be met by the systems available.

Visit Computer Installations

Set up visits to see computer maintenance programs in operation. Manufacturers can help set up demonstrations. Use resources of colleagues at other plants, conferences and trade shows to find out first hand how the systems operate before you decide to purchase.

Microcomputer Applications[1]

Not too many years ago, the personal computer (PC) was a novelty tinkered with by a few and ignored by many. In a relatively short period of time, however, this expensive and complex toy has evolved into an influential tool, asserting itself as one of the leading productivity enhancements available to industry.

---

[1] Reprinted with permission from the June 27, 1985 issue of *Plant Engineering* Magazine. Copyright 1985 by Technical Publishing, a company of The Dun & Bradstreet Corporation.

Today microcomputers have more processing power, more internal memory, and more storage capacity than most mainframes contained just a few years ago. With this development in hardware has come similarly dramatic, though perhaps not as visible, progress in software. Sophisticated, creative programming has led to truly user-friendly, menu-driven, software that is relatively low in cost and easy for even the novice computer user to understand and operate without extensive training. And the future holds the promise of even more progress, offering greater speed and efficiency, more programs, better graphics, and more powerful features.

The emergence of the personal computer has opened a powerful new world to the maintenance engineer. Now reasonable in cost and reasonably easy to use, microcomputers have made sizeable inroads into the maintenance engineering environment, with keyboards, monitors, and floppy diskettes replacing pencil and paper, calculators, file cabinets, and even drafting tables in the office and on the plant floor.

Along with the benefits, however, the PC poses some new problems. To be effective, indeed to operate at all, PCs must be equipped with appropriate, well-chosen software. Although programs can perform innumerable tasks—from the more general, all-purpose, data-base management functions and spreadsheet analyses to such engineering-related concerns as process monitoring and control; HVAC system, piping network, lighting, and wiring design and analysis; and maintenance management—successful application of software requires an investment in time and money. Suitable programs must be located, evaluated, and correctly applied to each defined task.

Figure 11-1 presents a comprehensive, though certainly not all-inclusive, listing of the most popular problem-solving programs available for the maintenance engineering field today.

It is designed to provide the reader with a concise summary of each program and to allow him to determine quickly system requirements, major features, and the cost of the packages that interest him. Each program description includes the address and phone number of the developer.

It is hoped that, using this directory as a starting point in his search, the maintenance engineer will be able to make his microcomputer a more viable tool and begin to build a software library that will allow him to respond more effectively to the dynamic, increasingly complex field in which he works.

CONCLUSION

There are presently a number of companies who sell maintenance and energy management software. Unfortunately, there is usually no "canned" software that can be used that will do all your maintenance work automatically. In every case you must supply the data that will allow the program to function as you want it to for your particular physical plant.

Required Data and This Book

You must provide location and identification data on each maintenance-dependent item. (See chapter 3 starting on page 13.)

You must provide the inspection and maintenance functions you want done. (See chapter 6 starting on page 27.)

You must provide the schedule you want to maintain. (See chapter 7 starting on page 33.)

You must specify the records and reports you want to maintain and or files. (See chapter 9 starting on page 39.) This chapter also indicates those items you want to produce for viewing or printing on hard copy.

You must specify the energy management records and reports you want to maintain or generate. (See chapter 10 starting on page 43.)

Only after you have completed this book and understand its principles can you begin to seriously consider a computer system.

Why Do Many Attempts At Computerizing Fail?

There are many reasons why computer systems fail. Chief culprits are in not understanding the original goals of the program, poor training and resistance to change. There are many complaints that have been voiced about computer maintenance programs including:

1. This system is nothing more than a glorified time clock.
2. This system generates more paper than we can read, let alone do the work.
3. This system is far too complicated for the average maintenance clerk to operate.
4. With this system you have to pay too much attention to every detail such as determining the manhours and prices of each part or item of material used on each job order.
5. This system was supposed to replace some of the maintenance staff. Instead we had to hire a computer operator. We still have a lot of paperwork.

The main reason for failure can be identified as management was not properly prepared to computerize operations. The key to success is to educate yourself to know the strengths and limitations of computerization. Computers do manage data and produce "good" records which give the maintenance manager more time to get the job done.

Figure 11-1. MAINTENANCE MANAGEMENT SOFTWARE

ABC/MM performs planning, scheduling, and control of the industrial maintenance function. Comprehensive, modular software covers all aspects of corrective and preventive maintenance, automatically schedules work orders in order of priority and date needed according to available manpower by craft, and maintains accurate backlog control. Menu-driven system also provides job estimating standards by time slots and for repetitive jobs. Program is written in COBOL and is available for a variety of microcomputers including the IBM PC ST and AT and the HP-3000. It requires 256K bytes of RAM and a 10M byte hard disk. Source code is available; a warranty is provided. Price: $3950 to $59,000. ABC Management, 805 Dupont, Suite 3, Bellingham, WA 98225. Phone: (206) 451-1116.

(Continued on next page)

B&A 201 stores and analyzes information on required, authorized, committed, and expended funds. System creates records and retains transaction information to permit sophisticated cost control, tracking, and analysis. Software interacts with this company's maintenance managment packages. Program is written in COBOL and runs under PC DOS, MS DOX, CP/M, AppleDOS, and Unix on most microcomputers. It requires 320K bytes of RAM and a 10M byte hard disk. Source code is available. Two training classes and a lifetime guarantee are provided Price: $3000 to $20,000. Planned Maintenance Systems, Inc. 5707 Seminary Rd., Falls Church, VA 22041. Phone: (703) 931-8090.

CAMMS fully interactive microcomputer software system manages and automates maintenance. Package includes six modules that cover storeroom operation, purchasing, work-order planning/scheduling, repair history, costs, predictive maintenance analysis, preventive maintenance, and report generation. Real-time system has graphics capabilities; all data entry and file maintenance is done on-line through color screens. Program is written in COBOL and runs under MS DOS and PC DOS on the IBM PC XT and AT. It requires 128K bytes of RAM and a 10M byte hard disk. Six days of training are included in the price. A warranty is provided. Price: $17,000. Cord Associates, 922 N. 26th St., Allentown, PA 18104. Phone: (215) 433-5020.

CAMMS is an off-line system for accurately acquiring, managing, analyzing, and documenting machinery vibration data. Analysis methods include alarm checking, vibration severity, bearing frequency calculation, and long term trending. Data, acquired from a variety of sources such as spectrum analyzers and vibration meters, or from keyboard entry, may be displayed or plotted. Program is written in BASIC and PASCAL and runs on the HP 9000 Series 200 desktop computers. A megabyte of RAM is required. A limited warranty is provided. Price: $9000. Structural Measurement Systems, Inc., 645 River Oaks Parkway, San Jose, CA 95134. Phone: (408) 263-2200.

CAMS-II controls preventive, predictive, and repair maintenance on a variety of plant equipment requiring regular maintenance. System tracks historical records and keeps a perpetual inventory of replacement parts and materials. A variable initiator scheduling

(Continued on next page)

system enables work to be planned by run time, frequency, calendar periods, and unacceptable readings, and on demand. Modules for general facility work orders, construction project tracting, and employee records are included. Program is written in C and runs under MS-DOX and PC-DOS on the IBM PC. At least 384K bytes of RAM and a 10M byte hard disk are required. Price includes training. A warranty is supplied. Price: $4450. Creative Maintenance Systems, Inc., P.O. Box 41226, Jacksonville, FL 32203. Phone: (904) 355-2745.

CATS, computer-assisted troubleshooting system, indentifies all possible mechanical and electrical faults for specific equipment and eliminates the guesswork in machine repair. Output lists corrective procedures, replacement parts and stock numbers, and alerts the user to any needed visual aids. Print-out may be generated and carried to malfunctioning equipment to assist with repair. Company will help the user develop a fault analysis data base. Program runs under CP/M 86, MS-DOS, and PC-DOS and requires 64K bytes of RAM. A warranty is provided. Price: $3500. Scientific Management Techniques, Inc., 581 Boylston St., Boston, MA 02116. Phone: (617) 536-3933.

CMTS is a computerized time standards program that helps maintenance departments estimate how long it should take to complete certain maintenance jobs. The system uses benchmark and slotting techniques to determine standard times. Software promotes accurate scheduling, ongoing productivity measurement, and consistent backlog reporting. Program is written in BASIC under MS-DOS for the IBM PC XT and AT. It requires 128K bytes of RAM and a hard disk. An audio/visual training course is available. Price varies. Modern Management, Inc., 7301 Carmel Executive Park, Charlotte, NC 28226. Phone: (704) 542-6546.

Comac maintenance management system consists of five modules to meet all the requirements of maintenance departments for planning, scheduling, job control, costing, resource allocation, history analysis, and record keeping. Additional modules are available for performing defect analysis and predictive/condition-based maintenance, and for generating customized reports. An inventory/engineering stores package that can interface with the maintenance system or operate alone also is available. Program is written in compiled BASIC and runs under MS-DOS and PC-DOS on the IMB PC

(Continued on next page)

XT and AT. It requires 256K bytes of RAM and a 10M byte hard disk. Price: $10,000; inventory system: $3500. Comac Systems, Inc., 6290 Sunset Blvd., Suite 1126, Los Angeles, CA 90028. Phone: (213) 463-5635.

EMCO+ tracks anything that needs repair or maintenance. Real-time, interactive, six-system package covers security, personnel, vendors, inventory, purchase orders, and equipment maintenance. Number of items tracked is limited only by computer storage space. Program is written in COBOL and runs under MS DOS on the IBM PC. It requires 512K bytes of RAM and a 10M byte hard disk. A warranty and 30 days training are included. Price: $47,500. Demar Service, Inc., 23236 Lyons Ave., Suite 203, Newhall, CA 91321. Phone: (805) 255-1005.

EMS 201 management information and work scheduling and control system for equipment covers preventive, predictive, corrective, and condition-based maintenance. Package helps managers plan, budget, schedule, execute, monitor, and evaluate a total equipment maintenance program. Analyses on equipment reliability, ownership costs, and repair or replace are provided. Program is written in COBOL and runs under PC DOS, MS DOS, CP/M, Apple-DOS, and Unix on most microcomputers. It requires 320K bytes of RAM and a 10M byte hard disk. Source code is available. Two training classes and a lifetime guarantee are provided. Price: $5000 to $60,000. Planned Maintenance Systems, Inc., 5707 Seminary Rd., Falls Church, VA 22041. Phone: (703) 931-8090.

Equipment Management System tracks failures and maintenance activities for a wide range of machinery and equipment. Reports include comparisons of like types of equipment, operating costs, and cost per unit. Costs may be tracked for up to 50 subsystems for each class of equipment. Program is written in PASCAL and runs under PC DOS and MS DOS on the IBM PC. It requires 256K bytes of RAM and two disk drives. A 1 yr warranty is supplied. Price: $699. Springfield Controls Corp., P.O. Box 127, Doylestown, PA 18901. Phone: (215) 924-0270.

EquipPLAN maintains equipment lists and asset data. Software prints equipment lists by department, manufacturer, date purchased, etc. Gross and net asset values can be maintained by running yearly depreciation on the assets. Program is written in FoxBASE

(Continued on next page)

and runs under PC DOS, MS DOS, Unix, and AOS/VS on the IBM PC, TI-PC, and DG Desktop. It requires 192K bytes of RAM. Telephone support and a limited warranty are provided. Price: $150. DacorMFG, 13330 Bishop Rd., Box 269, Bowling Green, OH 43402. Phone: (419) 354-3981.

Esti-maint prepares labor and material cost estimates and operation work standards for almost any maintenance task. Software functions separately or in conjunction with the company's *Maintainatrol* system. Program is written in MBASIC and runs under CPM/86. It requires 256K bytes of RAM and a 10M byte hard disk. Telephone support and a warranty are provided. Price: $2495. Advanced Business Strategies, Inc., 6465 Monroe St., STE B, Sylvania, OH 43560. Phone: (419) 882-4285.

FAMTRAC facilities automated maintenance management tracking and control system controls and regulates the timing of maintenance repairs. Comprehensive system operates from a menu of 42 available displays and printed reports to manage all aspects of maintenance including planning and scheduling, work-order issue, cost analysis, preventive maintenance, and inventory control. Software runs under PC DOS on the IBM PC XT or AT and requires a 10M byte hard disk. Program uses dBASE II (requiring 128K bytes of RAM) or dBASE III (requiring 256K bytes of RAM). Price of the program includes 2 day training course; a 90 day warranty is supplied. Price: $6000 to $6500. Syska and Hennessy, Inc., Engineers, 11 W. 42nd St., New York, NY 10036. Phone: (212) 556-3310.

IMP, inventory management program, reduces inventory costs through better stock management. Software interacts with the company's *MMP,* maintenance management program, and assists with issuing parts, maintaining accurate stock levels, reserving parts required by PM work orders, and more. Three levels of approval are provided for maintaining purchasing and historical data on vendor performance and inventory costs. Program is written in COBOL and runs under PC-DOS 2.0 or higher on the IBM PC. It requires 512K bytes of RAM and a 30M byte hard disk. Package includes 3 days training and a 1 yr warranty. Price: $15,000. NUS Operating Services Corp., 910 Clopper Rd., Gaithersburg, MD 20878. Phone: (301) 258-2444.

INV 201 is a sophisticated materials management system that helps reduce inventory costs. Software automatically produces purchasing

(Continued on next page)

documents and processes all requirements for materials and parts whether stocked or not. Program is written in COBOL and runs under PC DOS, MS DOS, CP/M, AppleDOS, and Unix on most microcomputers. It requires 320K bytes of RAM and a 10M byte hard disk. Source code is available. Two training classes and a lifetime guarantee are provided. Price: $5000 to $30,000. Planned Maintenance Systems, Inc., 5707 Seminary Rd., Falls Church, VA 22041. Phone: (703) 931-8090.

Mainline consists of more than 200 integrated program modules that handle work orders, inventory control, resources, equipment, preventive maintenance, pre-planned procedures, equipment history, employee records, cost accounting, and purchasing. Program is written in FORTRAN and runs under PC DOS and MS DOS on the IBM PC XT. It requires 256K bytes of RAM; a hard disk is recommended. Source code is available; a warranty is provided. Price: $5000. Morrow Technologies Corp., 3026 Owen Dr., Antioch, TN 37013. Phone: (615) 793-3555.

Mainsaver provides complete control of maintenance costs and personnel activities through planning and scheduling, preventive maintenance, work-order generation, spare-parts inventory and purchase ordering, labor records, and budgeting. Software generates a variety of reports including a monthly management report, predefined daily reports, and numerous status reports on equipment, facilities, crafts, and costs. Program may be configured for the single user or as a multiuser management information system. Software is written in dBase II assembler and runs under MS-DOS, CP/M-86, and concurrent PC DOS on the IBM PC and ATT-3B series. A minimum of 256K bytes of RAM, a hard disk, and a printer are required. Source code is available. Installation and training are included. A warranty is provided. Price: $3000 to $27,800. J.B. Systems, Inc., 21600 Oxnard St., Suite 640, Woodland Hills, CA 91367. Phone: (818) 340-9430.

Main/Tracker, manages information and functions related to the maintenance of equipment and facilities. System monitors and controls maintenance activities and maximizes the use and control of manpower, special equipment, tools, and repair parts resources. Five subsystems cover planning and scheduling, work-order control, repair parts inventory management, repair parts purchasing management, and fixed asset management. Program is written in RPGII

(Continued on next page)

and runs under PC-DOS and on the IBM PC XT. It requires 64K bytes of RAM and a 10M byte hard disk. A 6 mo warranty is supplied. Price: $4000 to $9500. Elke Corp., 998 Zane Ave. N., Golden Valley, MN 55422. Phone: (612) 546-1779.

Maintainatrol comprehensive maintenance management system consists of five interactive modules: work-order request, preventive maintenance, job/time card summary, equipment history, inventory control, and purchasing. Easy-to-operate system produces a variety of reports including information on any task, employee, trade, part, or asset. Program is written in MBASIC and runs under CPM/86. It requires 256K bytes of RAM and a 10M byte hard disk. Telephone support and a warranty are provided. Price: $3995. Advanced Business Strategies, Inc., 6465 Monroe St., STE B, Sylvania, OH 43560. Phone: (419) 882-4285.

Maintenance Management computerizes preventive maintenance. Software includes a master equipment record, repair history file, work-order file, past due report, maintenance scheduling and reporting, spare parts inventory data, and vendor and parts ordering information. Package has multiuser capability. Program runs under PC DOS, MS DOS, and CP/M 86 on the IBM PC and numerous other microcomputers. It requires 192K bytes of RAM and a 10M byte hard disk. A warranty, telephone support, and two days training are provided. Price: $5995. Acme Visible Records, 1000 Allview Dr., Crozet, VA 22932. Phone: (800) 368-2077.

Maintenance Management assists in planning and scheduling equipment preventive maintenance using printed work orders. Menu-driven, multiuser system features equipment and task files, a work-order scheduling module, and a variety of reports, including individual repair histories and analysis of types of repairs, worker performance, etc. Program is written in Oasis BASIC and runs under the Oasis operating system on the Onyx. It requires 192 bytes of RAM, a 20M byte hard disk, and a printer. A 1 yr warranty is provided. Price: $25,000 to $45,000. Institute for Business and Industry, Inc., 1927 Bristol Pike, Bensalem, PA 19020. Phone: (215) 639-4660.

Maintenance Management System covers preventive maintenance and unscheduled work orders, controls parts, and generates a variety of customized reports. Menu-driven system uses function keys for easy data entry. Company performs a field survey of equipment,

(Continued on next page)

develops a PM response program tailored to plant needs, implements the system, and trains personnel. Program is written in COBOL and runs under MS DOS on the IBM PC XT. It requires 256K bytes of RAM and a 10M hard disk. A warranty is provided. Price: $20,000 to $100,000. Crothall Systems West, 333 Hegenberger Rd., Suite 208, Oakland, CA 94621. Phone: (415) 569-9441.

Maintenance Management System organizes the maintenance management procedures required for a successful preventive maintenance program. System tracks up to 1200 procedures on up to 500 pieces of equipment. Procedures are automatically scheduled and printed on the basis of run hours or number of days since completion. System also stores equipment records and work histories, and includes an inventory control package that accommodates up to 1600 parts. Program is written in BASICA and runs under PC DOS and AppleDOS 3.3 on the IBM PC and the Apple IIe. Apple versions require 64K bytes of RAM; IBM versions require 128K bytes of RAM. Two disk drives are also needed. Telephone support is provided. Price: $849. Jentech Controls, Inc., Route 1, Box 93, Gresham, WI 54128. Phone: (715) 787-3795.

Maintenance Management System integrates preventive maintenance and repairs with a management information system. Package generates preprinted PM work orders and collects PM cost and repair data on each piece of eqipment. Machine history may be developed. Reports summarize PM not completed, labor and material costs, man-hour accounting information, and more. Program is written in COBOL and runs under MS DOS 2.0 on the IBM PC XT. It requires 128K bytes of RAM and a 10M byte hard disk. Telephone support and a warranty are provided. Onsite training is supplied. Price: $11,490. The Stanwick Corp., 3661 E. Virginia Beach Blvd., Norfolk, VA 23502. Phone: (804) 855-8681.

Maintenance Management System (MMS) performs four basic maintenance functions: equipment history, preventive maintenance, work-order generation, and spare parts inventory control. Craft definition is also permitted. Designed for those with no previous computer experience, the package is easy to use and generates a variety of reports. Program is written in C and runs under PC DOS on the IBM PC. It requires 256K bytes of RAM, two disk drives, and a printer. A warranty and telephone support are provided. Price:

(Continued on next page)

$5995, includes one day of training. Unik Associates, 12545 W. Burleigh, Brookfield, WI 53005. Phone: (414) 782-5030.

Maintenance Pac organizes maintenance data on equipment, inventory, vendors, manufacturers, and employees. Comprehensive user-friendly facilities software generates work orders, monitors work-order histories, organizes parts reordering, and generates reports quickly and easily. Program is written in compiled BASIC and runs under MS-DOS on the IBM PC. At least 128K bytes of RAM and a 10M byte hard disk are required. Package includes 3 day training course. Price: $7500. Datastream Systems, 11 Regency Hills Dr., Greenville, SC 29607. Phone: (803) 292-1790.

Maintenance Planning and Control is a complete on-line package that features work-order planning, preventive maintenance scheduling and follow-up, mean time to failure tracking, asset reporting, and labor distribution. Work orders are generated automatically. System is designed for maintenance personnel with no previous computer experience. Program is written in BASICA and runs under MS DOS on the IBM PC. It requires 128K bytes of RAM, two disk drives, and a printer. A warranty is provided. Training is available for an additional fee. Price: $1634. Maintenance Control Systems, Inc., 7530 S. Gallup St., Littleton, CO 80120. Phone: (303) 798-3575.

Maintenance Series is a group of maintenance management programs that performs numerous functions including preventive maintenance, inventory control, work-order processing, scheduling, systems management, equipment history, purchase-order tracking, spare-parts management, and more. System has multiuser and multitasking capabilities. Programs are written in compiled BASIC and run under MS-DOS on the IBM PC. From 128K to 512K bytes of RAM are required, depending on the number of modules purchased. Training session available for an additional fee. A 90 day warranty is supplied. Price: $995 to $9495. Penton Software, Inc., 420 Lexington Ave., Suite 2846, New York, NY 10017. Phone: (212) 878-9600.

Maintenance Spare Parts Inventory Control helps reduce inventory costs by generating purchase orders and parts and vendor lists; performing inventory valuing; and monitoring overstocking. System operates independently or in conjunction with this company's maintenance planning and control system. Program is written in BASICA and runs under MS DOS on the IBM PC. It requires 128K

(Continued on next page)

bytes of RAM, two disk drives, and a printer. A warranty is provided. Training is available for an additional fee. Price: $1024. Maintenance Control Systems, Inc., 7530 S. Gallup St., Littleton, CO 80120. Phone: (303) 798-3575.

Maximo computerized plant/facilities maintenance system provides comprehensive features for work-order tracking; inventory control, scheduling, job planning, equipment history, and standard reports. Special features include a report writer, PM scheduled by calendar time, run time, or mileage, system security, and the automatic generation of purchase orders from inventory thresholds. Software comes with its own mouse for easy entry of commands. Program is written in C and runs under PC-DOS on the IBM PC AT. It requires 512K bytes of RAM and a 10M byte hard disk. A warranty is provided. Price: $25,000. Project Software and Development, Inc., 20 University Rd., Cambridge, MA 02138. Phone: (617) 661-1444.

MC-Master controls daily inventory transactions through four files. Item Master maintains item descriptions, codes, costs, use, and purchase-order information. Stockroom Master File tracks part locations, amounts on hand, maximum permitted, out-of-stock items, etc. Vendor Master File maintains data on issuing purchase orders, vendor performance, and supplier selection. Purchase Order File references each item on file by vendor, purchase-order number, account number, and date expected. Program is written in MBASIC, runs under CPM/86, and requires 256K bytes of RAM. Telephone support and a warranty are provided. Price: $995. Advanced Business Strategies, Inc., 6465 Monroe St., STE B, Sylvania, OH 43560. Phone: (419) 882-4285.

Micro-Emis equipment management system provides precise detail on the type and source of costs for every component, every facility, and every maintenance worker in the plant. Software helps managers determine where to cut costs and provides the facts and figures to support cost-cutting decisions. Program uses dBase II and runs under MS DOS, and CP/M-80 and 86 on the IBM PC XT. It requires 256K bytes of RAM. A 1 yr warranty is provided. Price: $8000 to $12,000. Intec Systems Inc., 400 Australian Ave., West Palm Beach, FL 33401-9990. Phone: (305) 832-3799.

Micro Maint programs handle preventive, scheduled and breakdown maintenance work orders and scheduling, parts inventory control

(Continued on next page)

that covers issues, receipts, and status reports, and equipment history providing total maintenance cost, labor, and materials for the current and prior years. The system can handle up to 3000 machines, up to 20,000 procedures, and 10,000 maintenance parts. Software is written in COBOL, runs under PC DOS on the IBM PC/ST, AT, 5531 Industrial Computer, and a number of other microcomputers. It requires 128K bytes of RAM and a 132 col. printer. Source code is available under certain conditions. A warranty and telephone support are provided. Price: $3750. Diagonal Data Corp. (formerly Vertimax), 2000 E. Edgewood Dr., Lakeland, FL 33803. Phone: (813) 666-2330.

Micro-SIMS establishes, organizes, and implements the elements necessary to perform preventive and corrective maintenance. Package focuses on equipment information management, work requests, work-order planning and generation, equipment history, and preventive maintenance scheduling. Software streamlines routine steps and organizes resources, schedules, and priorities needed to keep equipment properly maintained. Program is written in FORTRAN/C and runs under PC DOS and Xenix on the IBM PC. It requires 256K bytes of RAM and a 20M byte hard disk. Source code is available. A 90 day warranty is provided. Price: $4995. Energy Inc., P.O. Box 736, Idaho Falls, ID 83402. Phone: (208) 529-1000

MMP, maintenance management program, schedules and issues correction and preventive-maintenance work orders; maintains a maintenance history; computes equipment, personnel, and maintenance performance statistics; and issues reports. Work orders include instructions, precautions, and a list of tools and parts. Software features flexible planning and scheduling capabilities. Program is written in COBOL and runs under PC-DOS 2.0 or higher on the IBM PC. It requires 512K bytes of RAM and a 20M byte hard disk. Package includes 5 days training and a 1 yr warranty. Price: $25,000. NUS Operating Services Corp., 910 Clopper Rd., Gaithersburg, MD 20878. Phone: (301) 258-2444.

MMS, maintenance management system, is a menu-driven program designed to increase plant operating efficiency. Software assists in the scheduling of equipment maintenance intervals. Maintenance routines take the form of work orders. A comprehensive set of reports assists with the forecasting of parts use and cost and man-

(Continued on next page)

hours required for maintenance. Program is built on dBase II and runs under PC-DOS on the IBM PC. It requires 256K bytes of RAM. Price: $4500. DP Systems and Services, Inc., 2120 Pinecroft Rd., Greensboro, NC 27417. Phone: (919) 852-0455.

Modcam provides the maintenance manager with tools necessary to reduce the cost and improve the level of service of maintenance activities. Software consists of seven interactive modules: work-order tracking, inventory control, equipment history, nameplate tracking, preventive maintenance scheduling, job planning, and purchase-order tracking. System is suitable for plants with as few as 10 maintenance employees and as many as 1000. Program is available in BASIC and FORTRAN. It runs under MS-DOS, PC-DOS, Xenix, Oasis, and RTE-A for the IBM PC, HP 150, and the HP 1000. It requires 128K bytes of RAM and a hard disk. Program is supported by an audio/visual training course and an established user's group. A warranty is provided. Price: $20,000. Modern Management, Inc., 7301 Carmel Executive Park, Charlotte, NC 28226. Phone: (704) 542-6546.

MTS, maintenance tracking system, is designed for the small maintenance shop. Software consists of an equipment file and six functional modules. Work-order module tracks the progress and cost of work orders; costs are monitored through the posting of labor hours and issuing of stock items to work orders. Work-order status and backing reports are available. Equipment histories also are maintained. Program is written in BASIC and runs under MS DOS on the IBM PC XT or AT. It requires 256K bytes of RAM and a 10M byte hard disk. Source code may be negotiated. A maintenance agreement warranty may be obtained. Price: $3000 to $5000. Daniel International Corp., Daniel Building, Maintenance and Industrial Services Co., Greenville, SC 29602. Phone: (803) 298-3500.

Overtime Manning maintains overtime hours per work order and expense numbers for individual trade codes. Software allows detailed entry of data, then generates a report summarizing overtime by department, shift, expense or work-order number, trade, and required hours. Program is written in R/M COBOL and runs under MS-DOS, TRSDOS, Xenix, and CP/M on a variety of microcomputers including the IBM PC and TRS-80 II. It requires 64K bytes of RAM. Telephone support and a 60 day warranty are provided. Price: $600.

(Continued on next page)

Vision Computer Systems, 3801 Monarch Dr., Racine, WI 63406. Phone: (414) 552-5007.

PartPLAN maintains all types of parts data. Information may be sorted in different ways to print parts lists, monitor cost information, and cross-reference parts. Software accommodates numerous codes for entering parts by number, name, buyer, scheduler, etc. Program is written in FoxBASE and runs under PC DOS, MS DOX, Unix, and AOS/VS on the IBM PC, TI-PC, and DG Desktop. It requires 192K bytes of RAM. Telephone support and a limited warranty are provided. Price: $150. DacorMFG, 13330 Bishop Rd., Box 269, Bowling Green, OH 43402. Phone: (419) 354-3981.

PEM, Plant Equipment Maintenance, contains four programs—plant maintenance, materials management, purchasing, and equipment data—that interact or function independently. Software uses the Seek data-base manager to enhance rapid access of information and allow the system to be tailored easily to specific installation environments. Program is written in Prime Information and runs under Pick on the IBM-PC, Wicat, and other microcomputers using the Pick operating system. A minimum of 150K bytes of RAM and a 20M byte hard disk are required. A 90 day warranty is provided. Price: $10,000 to $70,000. General Physics Corp., 10650 Hickory Ridge Rd., Columbia, MD 21044. Phone: (301) 964-6265.

Planned Maintenance System manages the preventive maintenance function. System prints PM work orders, and tracks PM labor and materials and what PM was not completed and why. Any PM procedures may be entered for any type of equipment. Program is written in COBOL and runs under MS DOS 2.0 on the IBM PC XT. It requires 128K bytes of RAM and a 10M byte hard disk. Telephone support and a warranty are provided. Onsite training is supplied. Price: $3000. The Stanwick Corp., 3661 E. Virginia Beach Blvd., Norfolk, VA 23502. Phone: (804) 855-8681.

PlantPLAN maintains data on plant repair and maintenance activity. Daily or weekly repair activities, hours, and material and labor costs can be entered and sorted. A variety of reports may be generated. Program is written in FoxBASE and runs under PC DOS, MS DOX, Unix, and AOS/VS on the IBM PC, TI-PC, and DG Desktop. It requires 192K bytes of RAM. Telephone support and a limited

(Continued on next page)

warranty are provided. Price: $150. DacorMFG, 13330 Bishop Rd., Box 269, Bowling Green, OH 43402. Phone: (419) 354-3981.

PMS, preventive maintenance system, establishes PM tasks and schedules; prints weekly work orders, inventory lists, and future work schedules; maintains and retrieves historical data on PM tasks; tracks labor and material costs; and more. Menu-driven system is flexible and expandable. Program is written in complied BASIC and runs under PC DOS on the IBM PC. It requires 128K bytes of RAM. Source code is available. Price: $495. Josalli, Inc., P.O. Box 460, Enka, NC 28728. Phone: (704) 252-9146.

Preventive Maintenance Scheduling provides all the tools needed to establish a complete PM program. System generates inspection forms automatically on the date required and tracks date of last maintenance performed, unrepaired problems reported, repairs completed, incomplete work, and more. Program is written in R/M COBOL and runs under MS-DOS, TRSDOS, Xenix, and CP/M on a variety of microcomputers including the IBM PC and TRS-80 II. It requires 64K bytes of RAM and hard disk. Telephone support, 1 week onsite assistance, and a 60 day warranty and provided. Price: $7500. Vision Computer Systems, 3801 Monarch Dr., Racine, WI 53406. Phone: (414) 522-5007.

The Preventive Maintenance System ensures that maintenance of selected equipment will be done at regular intervals. Easy-to-use program helps organize, operate, and control a preventive maintenance program. Numerous reports are produced, including a summary of units to be serviced and activities to be performed. Program is written in BASIC and runs under PC-DOS and MS-DOS on the IBM PC. A minimum of 128K bytes of RAM and two disk drives are required. Source code is available. A 90 day warranty is provided. Price: $250. Omni Software Systems, Inc., 146 N. Broad St., Griffith, IN 46319. Phone: (219) 924-3522.

Probe III computerized maintenance management system guarantees an increase in production uptime, reduction of stockouts, and an improvement in worker productivity. System covers inventory and stockroom management, work-order planning and scheduling, equipment history, and purchasing. Multiuser system is written in Databus and runs under PC DOS on a network of IBM PCs. Storage and hard-

(Continued on next page)

ware requirements are dictated by plant needs. Source code may be negotiated. A 90 day warranty is provided; training is available for an additional fee. Price: $75,000. Efax Corp., 444 N. York Rd., Elmhurst, IL 60126. Phone: (312) 279-9292.

Reliability Support Package provides computerized support for plant equipment reliability programs such as dynamic inspection scheduling, data collection by electronic notebook, rapid data analysis, and graphic reporting. System modules include vibration, lubrication, steam trap monitoring, preventive maintenance, and equipment history maintenance. Program runs under MS DOS and Unix on the IBM PC and the Fortune 32:16 and requires 512K bytes of RAM and a 10M byte hard disk. A warranty and one day onsite training are provided. Price range: $9600 to $28,600, depending on number of modules purchased. Reliability Center, Inc., P.O. Box 1421, Hopewell, VA 23860. Phone: (804) 541-5631.

Repmain II maintenance management package is designed to reduce equipment downtime and control the maintenance workload. Menu driven, multisuer system schedules and generates work orders, develops asset repair history by work order, and performs a variety of purchasing and cost management functions including budget projections, work-order cost analysis, and on-line inventory valuation. Package is modular and interactive. Program is written in PROGRESS and runs under UNIX on a variety of systems including the IBM PC AT and the AT&T 3B. A minimum of 512K bytes of RAM and a 20M byte hard disk are required. Source code is available. A warranty is provided. Price: $10,000 to $40,000. DLSA Inc., Box 496W, Waquoit, MA 02536. Phone: (617) 540-7405.

Spare Parts Inventory Control manages the spare parts required for preventive maintenance and equipment repairs. Software prints a list of parts according to an inhouse numbering system. Parts may be cross-referenced to manufacturers' or vendors' part numbers, quantities on hand, on order, costs, etc. Key reports include a stock status, reorder information, inventory value, and excessive parts on hand. Program is written in COBOL and runs under MS DOS 2.0 on the IBM PC XT. It requires 128K bytes of RAM and a 10M byte hard disk. Telephone support and a warranty are provided. Onsite training is supplied. Price: $3000. The Stanwick Corp., 3661 E. Virginia Beach Blvd., Norfolk, VA 23502. Phone: (804) 855-8681.

(Continued on next page)

ST-III is a modular multiuser maintenance management program that offers on-line, real-time interactive, integrated data-base management for facility and plant equipment maintenance. Available modules include work-order control, equipment history and specifications, preventive maintenance, parts inventory and purchase-order control, and cost and management information reporting. Program is written in PICK basic and runs on any microcomputer using the PICK operating system. System requires a hard disk and 256K bytes of RAM. Source code is available. Support and training and a 1 yr warranty are supplied. Price: $10,000 to $40,000. Sigma Consulting Group, 12465 Lewis St., Suite 104, Garden Grove, CA 92640. Phone: (714) 971-9964.

TAS 201 employee timecard and labor costing program posts labor hours and costs to this company's maintenance packages. Software analyzes direct labor costs and maintains employee work histories. Program is written in COBOL and runs under PC DOS, MS DOS, CP/M, AppleDOS, and Unix on most microcomputers. It requires 320K bytes of RAM and a 10M byte hard disk. Source code is available. Two training classes and a lifetime guarantee are provided. Price: $3000 to $20,000. Planned Maintenance Systems, Inc., 5707 Seminary Rd., Falls Church, VA 22041. Phone: (703) 931-8090.

Taskmanager is a complete maintenance management system for scheduling preventive maintenance (PM) activities. Corrective maintenance (CM) work orders may be generated on demand. System also tracks both CM and PM histories. An inventory management module makes reorder quantity recommendations automatically. Program is written in BASIC and runs under MS-DOS and PC-DOS on the IBM PC XT and AT. It requires 256K bytes of RAM and a 10M byte hard disk. Source code is available; a 60 day warranty is provided. Training is offered for an additional fee. Price: $995/yr lease, $4995 license. Global Software Consultants, Inc., P.O. Box 15626, Minneapolis, MN 55415. Phone: (612) 757-2035.

TMM, integrated total maintenance management system, supports work-order processing, equipment history, preventive maintenance, and inventory control. On-line inquiry capabilities permit rapid selection and viewing of all data. More than 20 reports may be generated; system features on-line help facility, a full screen editor, and single key inputs. Program is written in APL and runs under PC

(Continued on next page)

DOS and MS DOS on the IBM PC. It requires 512K bytes of RAM, a hard disk, and a 132-column printer. Telephone support is provided. Price: $4500. Weiss Associates, Inc., TMM Systems Div., 127 Michael Dr., Red Bank, NJ 07701. Phone: (201) 530-1805.

TMS, total maintenance system, generates corrective and preventive maintenance work orders and produces a variety of management reports on work-load schedules and priorities, work center performance, labor utilization and productivity, special activity tracking, maintenance histories, budget variances, and more. Menu-driven, user-friendly system may be customized to the plant and its special characteristics. Program is written in C and runs under MS-DOS, PC-DOS, concurrent PC-DOS, MP/M 80 and 86, and Xenix on a variety of microcomputers including the IBM PC. A minimum of 128K bytes of RAM and a hard disk are required. Price: $8295. HRL Associates Inc., 2102-B Gallows Rd., Vienna, VA 22180. Phone: (703) 448-1442.

Turnaround Management System is a shutdown/turnaround estimating, planning, scheduling, and control tool designed to minimize down-time and manpower requirements during a shutdown. User friendly, menu-driven software provides a quick and reliable method for evaluating progress, forecasting activities, and minimizing costs. System generates numerous customized reports. Program runs under MS-DOS and PC-DOS on the IBM PC and requires 128K bytes of RAM, two disk drives, and a 132-column printer. One week training session and unlimited phone support are offered. A warranty is provided. Price: $45,000. Turnaround Planning Services, 823 Bradwell, Houston, TX 77062. Phone: (713) 488-8187.

UpkeepPLAN maintains data on asset repair and maintenance activity. Daily or weekly repair activity is entered and used to produce daily or weekly repair reports. Data are retained by asset for future use. System is suited for installing a preventive maintenance system. Program is written in FoxBASE and runs under PC DOS, MS DOS, Unix, and AOS/VS on the IBM PC, TI-PC, and DG Desktop. It requires 192K bytes of RAM. Telephone support and a limited warranty are provided. Price: $150. DacorMFG, 13330 Bishop Rd., Box 269, Bowling Green, OH 43402. Phone: (419) 354-3981.

WMS 201, work management system, covers three areas: it is a work-order management and control system for the maintenance of build-

(Continued on next page)

ings, grounds, and equipment; a space utilization database program; and a building and grounds management information program. Package can be used to describe a plant to any level of detail and will accommodate virtually any plant activity including painting, inspection, repair, pest control, safety, landscaping care, and more. Functions include planning, scheduling, budgeting, assignment, monitoring, and evaluating operations and activities. Program is written in COBOL and runs under PC DOS, MS DOS, CP/M, Apple-DOS, and Unix on most microcomputers. It requires 320K bytes of RAM and a 10M byte hard disk. Source code is available. Two training classes and a lifetime guarantee are provided. Price: $3500 to $30,000. Planned Maintenance Systems, Inc., 5707 Seminary Rd., Falls Church, VA 22041. Phone: (703) 931-8090.

Work Order System generates and analyzes work orders. User enters data, including time, materials, and costs. Software generates numerous reports on hours and materials, overcharged hours, and weekly labor distribution; it also includes a 14 wk moving calendar status report that determines estimated and actual start and completion dates and the work completed. Program is written in R/M COBOL and runs under MS-DOS, TRSDOS, Xenix, and CP/M on a variety of microcomputers including the IBM PC and TRS-80 II. It requires 64K bytes of RAM and two floppy disks. A hard disk is recommended. Telephone support and a 60 day warranty are provided. Price: $3000. Vision Computer Systems, 3801 Monarch Dr., Racine, WI 53406. Phone: (414) 552-5007.

A
Appendix A:
Forms to Accompany Step Three

a Planned Maintenance Program

INVENTORY _____
ITEM   (COMPANY)   UNION (YES) (NO)
                    HUMAN RESOURCES

Name	Job Position	Licensed Trade School	Years On Job	Years In Co.	Remarks

ONE STEP AT A TIME

FORM 1

a Planned Maintenance Program

INVENTORY
ITEM _____ HUMAN RESOURCES

ONE STEP AT A TIME

FORM 1-A

a Planned Maintenance Program

INVENTORY ———————————
          (COMPANY)
ITEM ————————— BUILDING GROUNDS —————————

NO.	LOCATION	USE	SQ FT	NO. FLRS PK SP

ONE STEP AT A TIME

FORM 2

a Planned Maintenance Program

INVENTORY _____
           (COMPANY)

ITEM ___MACHINERY – VEHICLES___   EQUIPMENT

NO.	DESCRIPTION AND LOCATION	USES Air–Water–Steam–Oil–Chw. Gas–Hydr. Fluid–Chemicals	DATE PURCH

ONE STEP AT A TIME

FORM 3

a Planned Maintenance Program

INVENTORY_____
            (COMPANY)
ITEM_____ DRAWINGS_____

NO.	DESCRIPTION	ARCHIVES	DATE	REMARKS

ONE STEP AT A TIME

FORM 4

a Planned Maintenance Program

INVENTORY _____
                    (COMPANY)

ITEM _____   **PUBLICATIONS** _____

NAME	DATE	OPR	MAINT	PARTS	LOCATION FILED	REMARKS

ONE STEP AT A TIME

FORM 5

a Planned Maintenance Program

Description of Item	ID No. / Tag. No.	Drawing Ref. No.	Location Number & Written	Maint. Sect. Assigned	Serves Area or Item	Disconnect or Valve Location	Remarks

BUILDING NO.　　NAME　　ITEM INFORMATION

ONE STEP AT A TIME

FORM 6

FORM 7

PREVENTATIVE   MAINTENANCE   FORM

DWG REF.	MAINT I.D.	DESCRIPTION

ONE STEP AT A TIME	P. M.	DWGN.	LOCATION	DWG. NO.
		SCALE		
		DATE		

FORM 8

a Planned Maintenance Program   DATA   GENERAL EQUIPMENT

DESCRIPTION OF ITEM:

MAJOR ITEM
SERIAL NO. _____   SPECIAL ATTENTION _____
MODEL _____

SEE BELOW FOR COMPONENT INFORMATION

LOCATION OF ITEM

ITEM SERVES (AREA) (OTHER ITEM) (ETC)

MAJOR REPAIR RECORD

DATE	DESCRIPTION

MAINT. ID NO. _____

TAG NO. _____

UNIT NO. _____

DATE INSTALLED _____ COST _____
MANUFACTURER _____
SUPPLIER & ADD. _____

PUBLICATIONS

NUMBER	DESCRIPTION

DIAGRAMS/DRAWINGS/ETC:

SPECIAL TEST/CALIBRATION REQUIRED

FORM 9 – FRONT

MOTOR INFORMATION

NAME	SER NO.	HP	VOLT	AMP	PH	HZ	RPM	FRAME	SF	TYPE	LUBE	USE

PUMP INFORMATION/FAN

NAME	SER NO.	CFM GPM	MODEL	SIZE	RPM	COUPLING	LUBE	USE

COMPRESSOR INFORMATION

NAME	SER NO.	MODEL	TYPE	TYPE RECEIVER	USE

CONTROLLER/STARTER/DISCONNECT/INFORMATION

NAME	SIZE	ITEM	COIL NO.	HTR NO.	CONTACT NO.	USE

EXPENDABLES: PAPER/ROLLS/CHARTS/INK/PENS/FUSES
BATTERIES/BELTS/COUPLINGS/FILTERS/LIGHTS/PHOTOCELLS/CHEMICALS/OIL/GREASE/
GAGES/METERS/VALVES/BRUSHES/BEARINGS

NAME	SIZE	ITEM	RANGE	TYPE	AMOUNT

FORM 9 – BACK

a Planned Maintenance Program

DATA   **MEDICAL ELECTRONICS**

DESCRIPTION OF ITEM:

MAINT. ID NO. _____

MAJOR ITEM

MODEL NO. _____   SPECIAL ATTENTION _____

TAG NO. _____

SERIAL NO. _____

UNIT NO. _____

SEE BELOW FOR COMPONENT INFORMATION

DATE INSTALLED _____ COST _____

LOCATION OF ITEM

MANUFACTURER _____

SUPPLIER _____

ITEM SERVES (AREA) (OTHER ITEM) (ETC)

SUPPLIER ADDRESS _____

PUBLICATIONS

MAJOR REPAIR RECORD

NUMBER   DESCRIPTION

DATE	DESCRIPTION

DIAGRAMS/DRAWINGS/ETC:

CALIBRATION REQ/PERIOD/STANDARD/

GROUND FAULT TEST (FREQ)

IS THIS AN OSCILLATOR THAT REQUIRES FED OR STATE REGISTRATION?
(YES)   (NO)

FORM 10 — FRONT

MOTOR INFORMATION

NAME	SER NO.	HP	VOLT	AMP	PH	HZ	RPM	FRAME	SF	TYPE	LUBE	USE

PUMP INFORMATION

NAME	SER NO.	CFM GPM	MODEL	SIZE	RPM	COUPLING	LUBE	USE

COMPONENT MODULES/SUBASSEMBLIES/SUBCHASSIS

NAME	SER NO.	MODEL	USE

EXPENDABLES SUCH AS PAPER (ROLLS) (CHARTS) INK/PENS
BATTERIES/BELTS/COUPLINGS/FILTERS/LIGHTS/PHOTOCELLS/CHEMICALS/OIL/GREASE/
GAGES/METERS/VALVES/BRUSHES/BEARINGS

NAME	SIZE	ITEM	RANGE	TYPE	AMOUNT	USE

FORM 10 – BACK

a Planned Maintenance Program

DATA　　　　**BOILER EQUIPMENT**

DESCRIPTION OF ITEM:

MAJOR ITEM　　　　SPECIAL ATTENTION:

SERIAL NO. _____　SAFETY VALVE SET _____ PSIG

MODEL　　　　　　OUTSIDE THE BOILER ROOM

　　　　　　　　FUEL ELEC SHUTOFF SWITCH LOCATED

MAINT. ID NO. _____

TAG NO. _____

UNIT NO. _____

SEE BELOW FOR COMPONENT INFORMATION

LOCATION OF ITEM

BUILDINGS AND AREAS SERVED

BURNER INFORMATION:

DATE INSTALLED _____ COST _____

MANUFACTURER _____

SUPPLIER _____

SUPPLIER ADDRESS _____

MAJOR REPAIR RECORD

DATE	DESCRIPTION

PUBLICATIONS

NUMBER	DESCRIPTION

DIAGRAMS/DRAWINGS/ETC:

SPECIAL TEST/CALIBRATION REQ

ANNUAL INSURANCE INSPECTION

ACCUMULATOR TEST OF SAFETY VALVE

BOMB TEST OF FUEL TANKS

BLOWDOWN OF BOILER AND CHEM. TEST

BLOWDOWN OF WATER COLUMN GAUGE

SOOT BLOWING & STCK MON. LENS CLEAN

FORM 11 — FRONT

(ATTACH A COMPLETE VALVE LIST)

MOTOR INFORMATION

NAME	SER NO.	HP	VOLT	AMP	PH	HZ	RPM	FRAME	SF	TYPE	LUBE	USE

PUMP INFORMATION

NAME	SER NO.	CFM GPM	MODEL	SIZE	RPM	COUPLING	LUBE	USE

COMPRESSOR INFORMATION

NAME	SER NO.	MODEL	TYPE	TYPE RECEIVER	USE

CONTROLLER/STARTER/DISCONNECT/INFORMATION

NAME	SIZE	ITEM	COIL NO.	HTR NO.	CONTACT NO.	USE

EXPENDABLES: PAPER/ROLLS/CHARTS/INK/PENS/FUSES
BATTERIES/BELTS/COUPLINGS/FILTERS/LIGHTS/PHOTOCELLS/CHEMICALS/OIL/GREASE/
GAGES/METERS/VALVES/BRUSHES/BEARINGS

NAME	SIZE	RANGE	ITEM	TYPE	AMOUNT	USE

FORM 11 – BACK

a Planned Maintenance Program

BOILER INFORMATION SHEET						VALVES THERMOMETERS GAGES		
VALVES	USE	SIZE	TYPE	BRAND	DISC SEAT	PACKING	LUBRICATION	

THERMOMETER	USE	BRAND	SIZE	TYPE	SCALE

GAUGE	USE	BRAND	SIZE	TYPE	SCALE

FORM 12

a Planned Maintenance Program

BOILER INFORMATION SHEET					
FUEL TANKS Above ground with dike/submerged	Capacity	Type Fuel	Level Indicator	Type Heater	Remarks
TANK NO. 1					
TANK NO. 2					
TANK NO. 3					

FEEDWATER SYSTEM CLOSED SURFACE HEATER			
OPEN DEAERATING HEATER			
INJECTION STEAM HEATER			
DEAERATING TANK With trays & spray nozzles	Float and Level Controls	Usual Temperature 220-230°F	Pressure 4-5 lbs
VENT CONDENSER			
CHEMICAL TEST KIT			
SOOT BLOWER			
BLOWDOWN TANK			
ENERGY RECOVERY SYSTEM			
WATER COLUMN	GAUGE GLASS		
WARNING FLOAT			
TRICOCK			
BLOWDOWN VALVES (PETCOCKS)			

DETECTORS FLUE (BOILER VENT) THERMOMETER	"High Temp Indicates Dirty Tubes"
STACK DETECTORS	
FUEL IN (FEED) GAUGES (OVER STRAINERS)	
RECORDER	
ALL FIRE EXTINGUISHERS A-B-C TYPES	

FORM 13

a Planned Maintenance Program

						MAINTENANCE		
	PM		Bldg		PM			
Item	I.D.	Tag	No.	Location	Dwg. Ref.	Sect.	Sched	Priority

IDENTIFICATION NUMBER ASSIGNMENT

ONE STEP AT A TIME

FORM 14

a Planned Maintenance Program

		REFERENCE LIST		
NO.	GENERAL SUBJECT	TITLE	AUTHOR	PUBLISHER
		ONE STEP AT A TIME		

FORM 15

a Planned Maintenance Program

REFERENCE LIST			CONTROL SHEET	
NO.	ITEM	DATE/TIME OUT	DATE/TIME RETURN	NAME

ONE STEP AT A TIME

FORM 16

a Planned Maintenance Program

REFERENCE LIST			
NO.	GENERAL SUBJECT	TITLE/PAM NO. ETC.	LOCATION FILED

ONE STEP AT A TIME

FORM 17

a Planned Maintenance Program

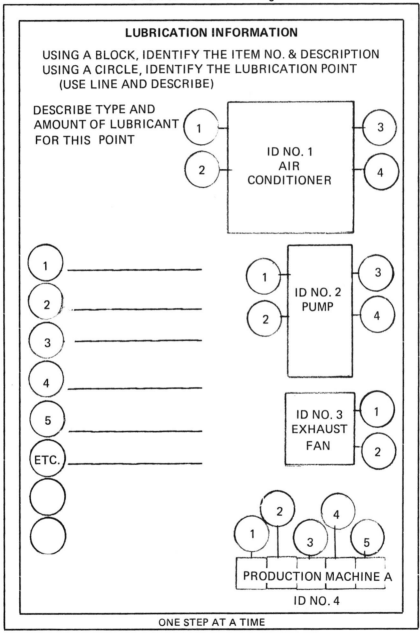

LUBRICATION INFORMATION

USING A BLOCK, IDENTIFY THE ITEM NO. & DESCRIPTION
USING A CIRCLE, IDENTIFY THE LUBRICATION POINT
(USE LINE AND DESCRIBE)

DESCRIBE TYPE AND
AMOUNT OF LUBRICANT
FOR THIS POINT

ID NO. 1
AIR
CONDITIONER

ID NO. 2
PUMP

ID NO. 3
EXHAUST
FAN

PRODUCTION MACHINE A
ID NO. 4

ONE STEP AT A TIME

a Planned Maintenance Program

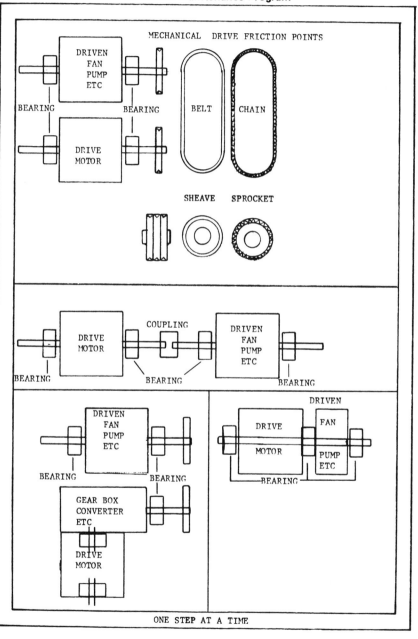

B
Appendix B:
Forms to Accompany Step Six

a Planned Maintenance Program

PREVENTIVE MAINTENANCE   ROUNDS

I.D. NO.	NAME	LOCATION		INSPECT					DATE
MAINTENANCE	DESCRIPTION	PHYSICAL	OVERALL	INDICATOR	FILTER	MOTORS	CONTROL	AREA	FROM TO
									DAY MO TU WE TH FR SA SU

ONE STEP AT A TIME

FORM 18

ACTION NEEDED OR TAKEN
INSPECTION FOLLOW UP   ☐ ✓ ONE
EQUIPMENT FAILURE   ☐

DATE _____

PREVENTATIVE MAINTENANCE FORM

A.
NAME OF ITEM

B.
LOCATION   INSPECTION   FREQUENCY

C.
MAINT. I.D. OF ITEM   J.O. No.

D.
ACTION REQUIRED

E.
PARTS USED   USE THIS SPACE FOR
                                              DIAGRAMS OR EXPLANATIONS

F.
PARTS NEEDED

G. ITEM   REPAIRED
        (IS) (IS NOT)

H. ITEM   IN OPERATION
        (IS) (IS NOT)

I.
NAME OF SUPERVISOR INFORMED

J.
MANHOURS   USED   (IN MINUTES ONLY)

K.
(REPAIRMANS) NAME (INSPECTOR)

L. SEND THIS FORM TO P.M. OFFICE

ONE STEP AT A TIME   **P.M.**

FORM 19

a Planned Maintenance Program

ROUNDS GENERAL INSPECTION

1. Item Operation: Normal, Irregular, Nonoperating
2. Indicators: Gauges, Lights, Flag, Timers, Charts, Correct or Erroneous
3. Controls: Switches, Breakers, Valves, Set Correctly, Intact.
4. Filters, Drains, Strainers: Drain, Clean or Change
5. Lubricants: Check, Fill as necessary.
6. Operators: Timers, Photocells, Floats, Operate Controlled Item Momentarily (If instructed to do so)
7. Readings: Record any required indicator readings, Change charts, Add ink, Clean as required.
8. Special Emphasis On: All items automatically controlled and all major systems or special devices that are manually controlled. Insure the item starts and stops and remains in the correct condition for that period it is required to do so (extra effort must be made in this area any time there is a power outage that could affect cycling, timing of a timer or restarting of motors, heaters, etc.).

NOTE: All Auto or Special Interest Items should be furnished. Rounds personnel on a list.

```
+-----------------------+
|                       |
|        ITEM           |
|                       |
+-----------------------+
```

☐ Complete, No missing or loose parts
☐ No Missing Insulation
☐ No Leaks
☐ No Unusual Sounds
☐ No Obstructions

ONE STEP AT A TIME

INSPECTION NO. 1

a Planned Maintenance Program

GENERAL AREA INSPECTION

ITEM	*INSPECT FOR*
SIGNS	Hung, Fixed and Positioned Correctly. Good State of Repair, No Loose or Missing Parts, Lettering is Legible, No Overdrawing or Graffiti, Not Outdated (No longer correct due to change)
LIGHTS	No Missing or burnt-out bulbs. Bulbs are correct wattage and type (for fixture and application). Fixture in good repair, wire, metal, glass, etc. No missing parts, not excessively dirty.
SPOT AND OUTDOOR	Properly positioned (Flooding Correct Area). Operating correctly, time on and off by timer, photocell, etc. No accumulation of water, snowbird droppings, etc. in or on fixture.
PLUMBING	All roof, floor and outside drains, clean, clear of debris, in good repair. No parts loose, damaged or missing. Valves tagged, media flow direction indicated, type of medium inside pipes properly secured. No leaks at valves, joints, controls, traps, pump systems and no sign of corrosion.
HVAC	Good Repair, Operating Properly. Filters clean, no piping leaks, no frayed wiring, no loose, damaged or missing parts. All indicators and controls functional and indicating properly.
AREAWAYS	Clean, clear of debris, properly marked and guarded.
LOADING DOCKS, DOCKS, WALKWAYS, ROADWAYS, PARKING AREAS	Clean, clear of debris, properly marked. All surfaces in good repair, properly illuminated and guarded.
ROOFS, CEILINGS, WALLS, FLOORS, DOORS, WINDOWS, SCREENS	In good repair. No sign of deterioration, discoloration, or corrosion. All moveable parts functioning correctly. No loose, damaged or missing parts.
FIRE AND EMERGENCY EQUIPMENT	Properly marked, stored and in a good state of repair, required test performed on time. No loose, damaged, missing or obsolete parts or items.
ALARM SYSTEMS AND DEVICES	In good state of repair. Properly functioning and set correctly for use assigned.

ONE STEP AT A TIME

INSPECTION NO. 2

a Planned Maintenance Program

THREE SENSE INSPECTION

1. SOUND
   Unusual Noise
   Banging
   Dripping
   Grinding
   Hissing
   Humming
   Slapping
   Silence

2. SIGHT
   Corrosion
   Deterioration
   Electrical Sparks, Wiring Faults
   Foreign Objects Obstructing Unit (Impairing Operation)
   Frayed, Loose or Missing Belts, Chains or Couplings
   Leaks, Air, Fluid or Lubricant
   Loose or Missing Parts
   Indicator, Gages, Thermometers, etc.: Intact and Set Correctly
       and Indicating Properly

3. FEEL
   Is Area or Item Hot, Cold, Damp, Wet, Drafty, etc.
   Are Bearings Abnormally Hot.
   Is Air Leaking from Closures.
   Is Air Moving in Ducts, Chambers, Plenums, etc.

ONE STEP AT A TIME

INSPECTION NO. 3

a Planned Maintenance Program

MAINTENANCE WORK ORDER   NUMBER _____

DATE _____ TIME/RCVD _____ DISPATCHER _____

CALLER _____ DEPT _____ TEL. NO. _____

DESCRIPTION OF WORK

WHAT

WHERE

WHEN IS START REQUESTED   (ACTUAL START)   (FINISH)

CRAFTSMEN ASSIGNED (NAME)

JOB COMPL. ACCEPTION (NAME)   (CUST.)   (MAINT SUPVR)

MANHOURS USED BY RATE _____

MATERIAL LIST AND COST _____

TOTAL COST _____

REMARKS: _____

SPECIAL PERMITS _____

SPECIAL TOOLS _____

REFERENCE LIST _____

UNUSUAL SAFETY CONDITION _____

NOTE: THE TWO-PIECE DAILY JOB ORDER AND HOW IT IS USED. THE TOP SECTION (ABOVE THE DOUBLE LINE) SHOULD BE A TEAR OFF SECOND PAGE TO BE ISSUED TO THE WORKMAN PERFORMING THE WORK. THE FULL PAGE (FIRST PAGE) SHOULD BE RETAINED BY THE DISPATCHER UNTIL THE JOB IS COMPLETED THEN THE MAN-HOURS BY RATE & MATERIAL COST MUST BE ADDED AND THE TOTAL COST OF THE JOB LISTED. THE SMALL FORM SHOULD BE FILED BY THE MAINTENANCE CLERK UNDER THE BUILDING AND DATE. THE SECOND (LARGER FORM) IS USED BY THE MAINTE-NANCE CLERK TO DO BUDGET WORK AND FILE.

(SEE NEXT TWO PAGES)

ONE STEP AT A TIME

FORM 20

MAINTENANCE WORK ORDER NO._____

DATE_____ TIME/RECD _____ DISPATCHER _____

CALLER_____ DEPT. _____ TEL.NO. _____
DESCRIPTION OF WORK

WHAT

WHERE

WHEN IS START REQUESTED   (ACTUAL START)   (FINISH)

CRAFTSMEN ASSIGNED (NAME)

JOB COMPLETION ACCEPTION (NAME) (CUSTOMER)   (MAINT. SUPERVISOR)

MANHOURS USED BY RATE _____

MATERIAL LIST AND COST_____

TOTAL COST

REMARKS:
SPECIAL PERMITS
SPECIAL TOOLS
REFERENCE LIST
UNUSUAL SAFETY CONDITION

ONE STEP AT A TIME

a Planned Maintenance Program

FORM 20—A, Page 1

MAINTENANCE WORK ORDER NO._____

DATE_____ TIME/RECD _____ DISPATCHER _____

CALLER_____ DEPT. _____ TEL.NO. _____

DESCRIPTION OF WORK

WHAT

WHERE

WHEN IS START REQUESTED   (ACTUAL START)   (FINISH)

CRAFTSMEN ASSIGNED (NAME)

JOB COMPLETION ACCEPTION (NAME) (CUSTOMER)   (MAINT. SUPERVISOR)

FORM 20–B, Page 2

a Planned Maintenance Program

WORK REQUEST / JOB ORDER / UNPLANNED NO._____

REQUESTER

DEPARTMENT _____ REQUESTER (NAME) _____ (TITLE) _____ (TEL. NO.) _____ (DATE OF REQUEST)

ITEM TO BE WORKED ON _____

LOCATION OF ITEM _____
(BUILDING)(FLOOR)(ROOM)(PARKING LOT)(ETC)

DESCRIPTION OF WORK

LABOR COST _____

MATERIAL COST _____

JOB COST (TOTAL) _____

LIST DRAWINGS
REFERENCES:
COMMENTS

WORK FLOW	
DATE ASSIGNED	DAILY (DO NOW)
	SCHEDULED
	DEFERRED (HOLD)
	CONTRACTED

ACTUAL DATE
OF WORK

EST. HOURS	ACT. HOURS	STARTED — COMPLETED	
		CRAFT ASSIGNED	NO.
		CARPENTER	
		CONTROL TECH.	
		ELECTRICIAN	
		ELECTRONIC TECH.	
		GROUNDSMAN	
		H.V.A.C.	
		LOCKSMITH	
		LABOR/SERVICE	
		MACHINIST	
		MASON	
		MECHANIC EQUIP.	
		MECHANIC VEH	
		MILLWRIGHT	
		PAINTER	
		PLUMBER	
		WELDER	
		EQUIPMENT OPERATOR	
TOTAL		ACTUAL NAMES ON BACK OF FORM	

(SPECIAL TOOLS OR EQUIPMENT REQUIRED)

(PERMITS REQUIRED)

(COORDINATE WITH DEPT.)

(PERSON TO CONTACT)(TEL. NO.) (DATE) (CONTACTED)(YOUR NAME)

MATERIAL REQUIRED

IN STOCK	MUST ORDER E.T.A. P.O. NO.	LOCAL PICKUP

(CUSTOMER ACCEPTANCE) (NAME) _____ (DATE)

ONE STEP AT A TIME

FORM 21

a Planned Maintenance Program

PLANNED MAINTENANCE
JOB ORDER NO. _____

DEPT. REQUESTING	JOB (TITLE) REQUESTOR		DATE

COST CENTER NR.	MAINTENANCE SUPERVISOR	BLDG. VEH. NR. OR EQUIP.

ACCEPTANCE
JOB APPLICATION

(NAME)	(TITLE)	(DEPT.)	(DATE)

DESCRIPTION OF WORK: ANNUAL MAINT, FABRICATE, INSTALL, DEMOLISH, REMOVE, ETC:

MATERIAL LIST: GASKETS, FILTERS, VALVES, BELTS, ETC:
ITEM   PART NUMBER   PRICE QUANTITY   TOTAL

SPECIAL (PERMITS) OTHER, REQUIRED

DESCRIPTION

RECEIVED (YES)(NO) _____
NAME

SPECIAL TOOLS: TORQUE WRENCH, LIFT TRUCK, POWER TOOL, ETC:

CRAFT ASSIGNED	NO.
CARPENTER	
CONTROL TECH.	
ELECTRICIAN	
ELECTRONIC TECH.	
GROUNDSMAN	
H.V.A.C.	
LABOR/SERVICE	
LOCKSMITH	
MACHINIST	
MASON	
MECHANIC EQUIP.	
MECHANIC VEH.	
MILLWRIGHT	
PAINTER	
PLUMBER	
WELDER	
EQUIP. OPERATOR	

REFERENCE PUBLICATIONS AND DRAWINGS LIST:

MATERIAL COST	MANHOURS IN MINUTES	CRAFTSMAN ASSIGNED BY NAME

ONE STEP AT A TIME

FORM 22

a Planned Maintenance Program

PROJECT ORDER NO._____

DESCRIPTION OF WORK

DATE SUBMITTED _____
DATE APPROVED _____
DATE WORK TO START _____
EST. DATE OF COMPLETION _____

LOCATION OF WORK

BUILDER _____
　　　　IN HOUSE, CONTRACTOR

PROJECT MANAGER _____

PROJECT COMPLETION DATE_____
　　　　　　　　　　ACTUAL

PROJECT COST_____
　　　　　　FINAL

ONE STEP AT A TIME

FORM 23

a Planned Maintenance Program

INSPECTION AND P.M. INSTRUCTION
AIR COMPRESSOR

1. The entire unit, motor, compressor, receiver and mountings should be inspected for any sign of deterioration or corrosion. Insure all items are intact and in a good state of repair.
2. Intake air filter (3 mo.).
3. Compressor for air or oil leaks.
4. Tighten bolts.
5. Oil Level.
6. Oil Change (3 mo.).
7. Belt tension (alignment, condition).
8. Sheave wear.
9. Grease (repack motor annually).
10. Lubricate motor (oil) (3 mo.).
11. Check motor shaft for excess end play (3 mo.).
12. Cylinder relief valve.
13. After cooler (drain). Check water out temperature.
14. Safety relief valve, all check valves.
15. Drain receiver.
16. Electric wiring (voltage, amps).
17. Motor starter and terminal connections not loose or burnt.
18. Any strange knocks or rattles.
19. All piping, valves (hand, electrical or pneumatic).
20. Pressure control switch.
21. Gauges.
22. Area under and around compressor is clean, dry and clear of obstructions.

ONE STEP AT A TIME

INSPECTION NO. 4

a Planned Maintenance Program

INSPECTION AND P.M. INSTRUCTION
MOTOR: Without Disassembly

1. END BELLS
   A. Tight
   B. No bolts
2. SHAFT
   A. End play? If more than 1/8", report
3. BEARING
   A. Turn shaft, should not be stiff or noisy
   B. Properly lubricated
4. BRUSHES
   A. Broken
   B. Worn
   C. Missing
   D. Misaligned
   E. Connection Firm
   F. Holders
   G. Adjusters
   H. Tension—setting
5. MOUNTING
   A. Loose
   B. Broken
   C. Worn
   D. Musty
6. ELECTRICAL CONNECTIONS
   A. Firm
   B. No frayed wire
   C. No exposed connections
   D. No bare copper
7. CONDUIT
   A. Good, mechanical connection
   B. Properly secured
8. DIRTY
   A. Being dripped on
   B. Dusty Area
   C. Other
9. MEGGER

NOTE: PREVENTIVE MAINTENANCE ACTION
1. If dirty—oil or other contaminant removed with cleaner, brush, rag, etc.
2. Blow off outside with air blower. Inside with heated blower.
3. Tighten.
4. Replace missing or broken item.
5. Lubricate.
6. Report other maintenance for replacement.

ONE STEP AT A TIME

INSPECTION NO. 5

a Planned Maintenance Program

INSPECTION AND P.M. INSTRUCTION

DRIVE ITEMS

GUARD–SPROCKET AND CHAIN

1. Guard
   A. Remove Guard
   B. Inspect for wear or missing brackets, etc. Rust.

2. Sprocket
   A. Alignment with opposite sprocket
   B. Loose on shaft–shake with both hands. (Check bearing if used.)
   C. Loose hub–bolt–washer–nut (loose or missing)
   D. Key loose–missing–worn
   E. Teeth worn–broken–missing

3. Chain
   A. Link–cracked–worn–loose on pin
   B. Link pin–no locking ring–or end of pin rubbed off
   C. Master link–loose–lock missing
   D. Stretched–measure pitch
   E. Loose (Take-up–adjusted properly?)
   F. Take-up–missing bolt or other problem

4. Does this sprocket or chain have lub?

NOTE: PREVENTIVE MAINTENANCE

1. Lubricate
2. Alignment–correction
3. Item missing–replace
4. Item loose–tighten
5. Worn, stretched (report to maintenance for replacement)

ONE STEP AT A TIME

INSPECTION NO. 6

a Planned Maintenance Program

INSPECTION AND P.M. INSTRUCTION

BELT INSPECTION
1. Is belt stretched?
2. Is belt hard?
3. Is belt frayed?
4. Is belt shiny?
5. Is belt broken?
6. Is tension correct?
7. Is alignment correct?
8. Is belt sized correctly for this application?
9. Is belt type correct for this application?

SHEAVES INSPECTION
1. Are sheaves correct type and size, in good repair, not broken, worn, wobbling on shaft, loose on shaft or misaligned?
2. Are all keys and/or lock screws in place and properly secured?

ALIGNMENT INSPECTION (See figure.)
1. Align the fan and motor sheaves by using a straight edge. The straight edge must be long enough to span the distance between the outside edges of the sheaves. When the sheaves are aligned, the straight edge will touch both sheaves along their diameters. A string, drawn tight, may be used in the same manner. When the sheaves are not the same width, place the string in the center groove of both sheaves and pull tight. The string should go straight through the center of both grooves. With the sheaves aligned, tighten the sheaves to their shafts.
2. Place the belts over the sheaves being careful not to use excessive force which can rip the cords in the belts. A matched set of belts will provide a more uniform drive loading.
3. Back the motor off until all belts appear snug. Operate the motor for a few minutes to allow the belts to set properly.
4. Adjust belts to proper tension by resetting motor position. The diameter of an adjustable pitch must not be changed for purpose of adjusting belt tension. Belts must be tight enough to avoid slippage during operation. *Excessive tension, however, will shorten bearing life on the fan and motor will accelerate belt wear.*

(continued on next page)

ONE STEP AT A TIME

a Planned Maintenance Program

ALIGNMENT INSPECTION *(continued)*

5. Check the tension at least two times during the first day's operation. Normally, there will be a decrease in belt tension as the belts stretch during initial run-in.

6. *It is normal for belts to squeal slightly at start-up.* With across-the-line motor start, this might be one or two second duration on a ten second start-up.

7. On fans shipped with drive assembled, follow the above steps of alignment and belt tension.

ONE STEP AT A TIME

INSPECTION NO. 7

a Planned Maintenance Program

INSPECTION

PUMPS

The entire assembly—pump, motor, coupling and base—should be inspected for signs of corrosion or deterioration and to insure that the mounting, wiring and piping are intact and not loose, missing or frayed.

The coupling and guard are intact and fastened correctly.

Packing or seals are not worn or adjusted all the way in and out allowing water blow by.

If water cooled, insure water operates with pump.

Water drains under packing gland or water coolant are clean and clear.

Unit has proper type and amount of lubricator (motor, pump and coupling).

Piping into and out of the unit and all hangers and insulation in a good state of repair.

Electric wiring, conduit and controls are intact and in a good state of repair.

Listen for unusual noises from bearings, couplings, motors, impeller, disconnect.

Look for leaks—corrosion around fittings, connections, valves, etc.

Feel for excessive heat around bearings, motor and electrical controller.

If unit is belt-coupled (motor to pump) check sheave condition and alignment belt tension and condition.

Notice any indicating devices such as gauges, thermometers, recording charts, etc. Are they in good repair and indicating correctly?

NOTE: REMEMBER THE THREE SENSES CHECK!

ONE STEP AT A TIME

INSPECTION NO. 8

a Planned Maintenance Program

INSPECTION

CIRCULATING PUMP

Entire unit for wear, deterioration or corrosion, parts missing or loose.

MOTOR: Heat, noise, sparking and lubrication.

MOUNT HANGER: For loose or missing parts.

COUPLING: Alignment, wear and shaft security. Coupling guard in place.

IMPELLER: Move medium through piping.

HOUSING: Packing, mechanical seals and gaskets loose, worn, missing, and unit leaking.

WIRING: Broken, frayed, missing or loose.

PIPING: Leaking, loose, corroded or missing.

Valves leaking, don't close or open.

Indicator in good repair and indicating properly.

ONE STEP AT A TIME

INSPECTION NO. 9

a Planned Maintenance Program

INFORMATION REQUEST

AUTOMATIC CONTROLS: List all items that are controlled automatically by such items as timers, photocells, float switches, etc.

MANUAL CONTROLS: List all items that must be turned on or off manually at a given time day/week/month/seasonal/etc.

CONTROL ITEM	CONTROLS ITEM & LOCATION	LOCATION CONTROL DEVICE	CYCLE HOURLY/DAILY/ETC

ONE STEP AT A TIME

FORM 24

a Planned Maintenance Program

DETAILED INSPECTION — HVAC

Major Air Conditioning Systems Using Steam and Chilled Water

1. OVERALL

   The entire housing, casing and connected duct work should be inspected for any sign of corrosion or deterioration and to insure that hangers and mountings are intact. Vibration/Isolation devices are intact and functioning properly, access openings are properly secured with no holes (ruptures), no missing insulation, screws, bolts, handles and hinges. Duct and housing insulation in good serviceable condition neither torn, loose or missing. All guards are in place and secured. All intake and exhaust are clear. There are no leaks of water, steam, air or oil. No excess or spills of lubrication. All valve handles, sight glasses, gauges, and indicators of all types are intact, clean, clear, set properly and indicating correctly.

2. FRESH AIR IN

   A. Deflector/Damper (see next paragraph for damper)
      Blades intact, not missing, loose, bent or torn.
   B. Screen
      Intact, not missing, loose or torn. Clear of debris, leaves or paper. No outside cover (by mistake). Storage against unit, etc.
   C. Thermometer outside air.
      Bulb and indicator intact. Clean, clear, fastened securely and indicating properly.

3. DAMPER

   The blades should be intact not loose, bent or misaligned. No missing, loose or corroded linkage. The operator and fasteners should be intact, secure and operating correctly. (Special emphasis should be given to fire dampers.)

4. FILTERS

   This is the area that the good P.M. man will be very thorough with. The purpose of the filter is to provide the system with clean air. If the filter is clean, air passes through it with enough volume to have excellent air supply, but as the filter surface is dirtied it begins to restrict air flow and after the surface has reached a certain saturation the fresh air volume is inadequate for proper operation of the system.

   NOTE: Be careful in heavy snow; it will be drawn into the prefilters and cut off your air supply. The filter also keeps outside dirt and foreign material from entering the heating and cooling coils. This would develop a coating on the fans and coils. This then would lower your heating and cooling capacity. Also, any opening in the housing on the suction side of the fan will allow dirt or foreign material or the room air (hot or cold) to be drawn into the system.

(continued on next page)

ONE STEP AT A TIME

a Planned Maintenance Program

4. FILTERS *(continued)*
    When changing filters, first record the reading on the air flow indicator (Magnehelix or similar). TURN THE FAN OFF. This stops dirt from the filters being removed from entering the system. Does not allow air to enter system through filter opening. Does not draw indoor air through opening you entered or filter access cover you removed to reach filter.

    NOTE: Sometimes it is justifiable to replace filters with unit running, but this should be authorized by foreman or supervisor.

    When installing filter:
    Insure you have the correct filter. Check that filter is located properly and air flow direction correct. Install fasteners correctly. If filter is a roll-a-mat, jog system.

    After the filters have been changed and old filters removed and area cleaned, be very careful to close and get a good seal on the door or access cover before starting fan. Turn fan on record indicator reading (Magnehelix, etc.) and measure amperage of main fan motor. Most major air conditioning systems of this type have prefilters and filters. That is actually two filters, one in front of the other.

    The first one, or prefilter, is normally an inexpensive throwaway type that extends the useable life of the main filter which is usually expensive. There are many types of filters. A few are: pad throwaway; pad permanent reuseable; box (absolute type); bag; electric. Some of the filters have an automatic cleaning system such as fluid spray. *Be aware* that when working on an electronic filter there is extreme danger of shock if equipment is not turned off.

5. PREHEAT or HEATING COIL must be kept clean and clear with no bent fins; no corrosion of pipes, hangers, supports. No tubing leaks or corrosion. If the control is calling for heat, the automatic valves (pneumatic or electric) should be open and the steam trap on the outlet side should be hot to the touch.

6. FREEZSTAT
    This is a special thermostat device with a long bulb extending over an area of the plenum (normally) located just after the preheat coil. Its purpose is to stop the fan and close the fresh air damper should the temperature of the air passing through the preheat get low enough to freeze the coils. You can normally close off the fresh air and raise the setting on the preheat coil to operate with return air, or open the access ahead of the filters and allow room air to enter the system. Some freezstats have remote signal or light indicators.

(continued on next page)

a Planned Maintenance Program

7. HUMIDIFIER
   A. Water Spray Type
      On command, this system pumps water from a sump pump and sprays it over the cooling coils. The water runs down the coil and into the sump. The sump level is maintained by a float valve on the make-up water and an overflow drain to bleed a small amount of water to eliminate dissolved solids from the sump. This sump should be tested for bacteria buildup and algae growth. Chemicals should be added as necessary.
   B. Steam
      On command, this system causes steam to be blown into the discharge air duct. Be careful of condensate buildup in the duct.

8. COOLING COIL
   The cooling coils are employed to control the quantity of moisture in the air. As the air passes through, moisture condenses and runs down into a collector pan. The cooled air is then drawn through the fan and reheated as necessary. In most systems, chilled water in has temperature in and pressure indicators, and a temperature out indicator. These should be noted. Check hangers, supports and drains. Also check fins and tubing.

9. FAN
   The fan is nothing more than a pump propeller which has been selected to operate at a certain pressure, discharging a designed quantity of air on a continuous basis. Both the fan and motor mounting, bearings and housing must be in a good state of repair. The belt sheaves and belts should be in alignment. The belts should be in good condition. The starter must be operating correctly with all wiring and connectors and connections tight. If unit has timer, it should be functioning and set correctly.

10. REHEATS
    Reheats are employed to modify the discharge air temperature. These coils are small units similar to the heat coils and must be maintained the same way.

11. CONVECTORS
    This is the device that passes the air into the temperature controlled area from the ducting. Some are adjustable and some are fixed. The fins should be straight, properly aligned, operate freely if adjustable, and be free and clear of obstruction.

12. FIRE OR SMOKE SENSORS
    These units should be clean, clear of obstruction and in a good state of repair.

ONE STEP AT A TIME

INSPECTION NO. 10

a Planned Maintenance Program

P.M. INSTRUCTION — HVAC

Major air conditioning systems using steam and chilled water used in conjunction with the "detailed instruction for inspection of air conditioning systems of steam and chilled water."

1. OVERALL
   Tighten: loose parts, fittings, couplings, covers and guards.
   Replace: indicators, insulation (small area), handles, access covers, filters, lubrication, chemicals, belts, sheaves, couplings, hoses, wiring, switches, controls, motors, pumps, spray systems (water and steam), electronic filter parts such as wires and hangers, valves and steam traps as well as fuses and starter heaters.

2. FRESH AIR IN DEFLECTORS
   Tighten, straighten and replace as necessary. Fresh air in screens that keep out birds, trash and foreign objects: tighten, mend, clean or replace as necessary.

3. FRESH AIR DAMPER
   Lubricate, tighten, adjust blades to linkage or handle, if auto, check linkage to operators, check operator for air or hydraulic (oil) leaks and insure hanger (fastener is tight and secure). If the damper blades are set properly to the linkage and linkage to the operator, the blades should close off (seal) and open wide when operated. When set properly, scribe mark on linkage at correct position. To test blades and mechanism, disconnect operator from linkage and operate by hand. If tight, add slight amount of penetrating oil at pivot points (bearings).

4. FILTERS
   Should be the correct size and type for the equipment! Should be clean, in proper position (with correct air flow direction), not broken, bent or distorted and properly fastened. Inspect them and replace when dirty or when air flow indicator (Magnehelix, Hays, Dyer, etc.) has reading specified for filter replacement, or at interval specified by instructions.

   Changing the Filter: If the unit has air flow indicator, you should record the final (dirty) reading. Turn fan off, change filters, insure correct size and type of filter, air flow direction, and that it is seated correctly and secured. Remove old filters, clean area, replace access covers or doors and assure a good air seal and that there are no lost or loose screws, bolts, handles or hinges. Start fan and take clean filter air reading and record it.

   A. Electronic Filters
      When any work is to be performed on the filter, you must insure that the power supply to the filter is turned off—*extreme personal danger exists at this unit.* Filter should be clean and in good repair. Inspect for dirt buildup; clean as necessary. Check for broken wires

(continued on next page)

ONE STEP AT A TIME

a Planned Maintenance Program

FILTERS—Electronic *(continued)*
and replace any that are missing. Turn unit on and the circuit breakers should stay in and the indicator lights should indicate properly.
If not, turn unit off and look for broken wires or a foreign object in
the filter or any wiring or fixture short.

B.   If unit has a wash unit, check mounting, hoses, fittings, nozzles,
drive fluid supply, the drain system and the control system. If the
unit has a bacteria or algae killer chemical agent: If *manual* test and
change as necessary; if *automatic* test and add chemical to reservoir
as necessary. Check pump and motor for lubrication, coupling, electric wiring and controls.

C.   *NOTE:* If return air is filtered, insure it does not have a contaminant
that could be harmful to you. If you do change a contaminated
filter, be sure to wear the proper protective clothing and mask, and
follow special instructions for disposal of filters.

5.   HEAT COILS
Inspect coils and fans for leaks and corrosion. Repair loose or broken
hangers. Inspect automatic control valve on steam line in for leaks if
pneumatic for air and steam; if *electric,* for wiring and steam leaks. If the
unit is calling for heat, the valve should be open and the piping in, the
coil and steam trap on the output pipe should be hot. If not, check auto
valve position, the associated hand valve, drain the strainer ahead of the
steam trap, and finally check the trap itself, but "first" check the feed
pipe to the main auto control valve to insure steam is being sent in on the
mains to this unit. Steam Trap Check: Insure the trap is not fouled by
excessive back pressure.

6.   FREEZESTAT
Inspect bulb and hangers. Check temperature setting. Insure control is
fastened securely and in good state of repair, no parts missing or loose
and the item is not obstructed! Check that it is properly marked and if it
has remote indicator, insure it is operational.

7.   HUMIDIFIER
A.   Water Spray Type
Insure water level in sump is correct; some water should be bleeding
to help stop solids from building up in the sump and being sprayed
onto the coils by the pump. Operate the float switch and insure fresh
water enters and stops when float is released. Adjust as necessary.
Check pump suction line filter screen, clean or replace it as necessary. Check pump operation; turn it on and check all piping and insure all spray nozzles are operating and directed correctly onto coils.
Clean heads, clean and tighten pipe fittings, check for loose, frayed
or broken wiring on the drive motor and pump, lubricate them as
necessary. If the sump is dirty, drain, dry and clean it and insure
drain system is clean and clear. Test for bacteria buildup or algae
growth. Apply chemicals as necessary.   *(continued on next page)*

a Planned Maintenance Program

HUMIDIFIERS *(continued)*

B. Steam Type Humidifier

Check area of ducting for excessive moisture. Is steam on continuously, is humistat calibrated correctly. Check humidifier housing. Clean and operate control. Insure steam injector (pipe hole or spray orifice) or pot are open. Check steam line, filter, screen drain. Clean or replace as necessary. Check steam trap; it should be hot when you operate the unit. If not, drain the strainer and operate the unit again. If steam line to humidifier is hot, but trap is cold, turn off steam and check trap. See HEAT COIL—Item 5.

8. COOLING COILS

The cooling coils are employed to control the moisture quantity in the air. As the air passes through the coils, the moisture is condensed and runs down into a collecting pan. The cooled air is then passed through the fan and is reheated as necessary for the space served.

In most systems, the chilled water has a temperature and pressure indicator on the input line and a temperature indicator on the output. These readings should be noted. The cooling coils should be back flushed periodically. The coils should be inspected for leaks and deterioration and the fins should not be bent or missing. The drain system should be intact, clean and clear. If cooling coils are on suction side of the fan, careful attention must be paid to the draining of the collection pan, since it is possible to build a level of condensate in the pan equal to the number of inches of water at which the system is designed to operate.

9. FAN

RPM should be measured and coincide with correct reading. The motor amperage should be read and recorded as well as the RPM reading just mentioned. Insure the motor is in good condition, electrical connections in good repair, the bearings of both fan shaft and motor pass noise and heat test. For noise test, use assigned listening test equipment or place your ear against the end of the handle and listen (with a little practice, you will soon know a growling bearing when you hear it). Touch the bearing and determine the level of heat it is generating (remember some bearings run hot). You must know what is normal for this particular bearing in this application. The bearings should be repacked annually. Insure the belt pulleys (sheaves) are aligned and belts have proper tension. Check that the belts are in good condition. Check that the fan shaft is not bent and no unbalance of the fan is noted. Tighten hangers and mountings, adjust vibration/isolation devices. Be sure all access openings are secured and intact. Check the electrical starter and pneumatic controls for tight connections and completenesss, no unusual noises or sparking, air hissing or burnt insulation smells or condensate draining or corrosion on controls or tubing. Also be sure all covers are in place and all indicators are functioning properly. Check lights, fuses, heaters and audio indicators.

(continued on next page)

ONE STEP AT A TIME

a Planned Maintenance Program

10.   REHEATS

Reheats are small heat coils similar to the heat coils and must be checked the same way including control operation and trap on output.

Electric Reheat: On command, the contactor should pull in and the element should develop heat. Check wiring and connections for loose or missing fittings or connections. Check the control disconnect and thermostatic control for functioning, completeness.

11.   CONVECTORS

Insure they are clean and clear. These devices direct the air into the controlled space. If they are dirty, this indicates dirty filtering or open access allowing bypass of filters. The vanes should be in good condition, not missing, bent or distorted. Some are fixed and some are adjustable; the mechanism should operate freely with the proper tool.

ONE STEP AT A TIME

INSTRUCTION NO. 1

a Planned Maintenance Program

QUICK CHECKLIST FOR MAJOR AIR CONDITIONING SYSTEMS USING STEAM AND CHILLED WATER

1. The item identification tag is visible and in good repair.
2. There are no unusual sounds, smells or sights such as leaks, loose or missing parts.
3. All controls, electrical, steam, water and air are in good repair, properly set and indicating correctly.
4. All indicating devices such as gauges, thermostats, lights and audio (bells) etc. are clean, in good repair, properly set and indicating correctly.
5. All electrical wiring and devices are in a good state of repair. No frayed or broken wires, nothing loose or parts missing, no motors sparking or making noise.
6. The couplings on motors, pumps, fans, etc., are in good repair, properly aligned, tightened and guarded.
7. The belts and sheaves (pulleys) on motors, pumps, fans, etc. are in good repair, properly tensioned, aligned and guarded.
8. All coils, steam and water are in good repair and clean. No bent fins or leaking tubes.
9. All filters clean, in place, air flow correct and in good repair.
10. All dampers (vanes, linkage and operators) in good repair. Lubed, tight and clean. Set at correct position.
11. All lubrication is up to date.
12. All chemical additives and devices in good repair and properly filled.
13. The entire housing, casing and connecting ducts from fresh air to discharged air should be inspected to insure that hangers, mountings and insulation are intact. Vibration, isolation devices are in good condition and secure.
14. The piping, steam, water, chemical and air, with all fittings and valves in good repair, with no sign of leaks or corrosion. All valves properly set and properly insulated.
15. Water pans, sumps, drains clean, properly filled if empty.
16. Discharge temperature should be recorded.

ONE STEP AT A TIME

INSPECTION NO. 11

a Planned Maintenance Program

INSPECTION PROCEDURE: EXHAUST FANS (GENERAL)

DAILY
1. Feel air flow.
2. Listen for unusual noise.
3. Look for loose or missing parts.

SEMI-ANNUAL
1. Remove cover, inspect and clean as necessary, the air exhaust screen of dirt, trash or obstructions. Inspect damper, operator, blade and linkage. Inspect high temperature limit switch or smoke detector, if applicable.
2. Inspect belt tension, alignment and condition, adjust or replace as necessary (if belt is hard, broken, frayed or shiny—replace). If loose, adjust the motor mounting taking care to keep the proper alignment. Remember, belts do stretch, so insure the belt is still serviceable.
3. Inspect, clean and lubricate motor, inspect wiring, switch and conduit for loose or missing parts, and tighten any loose connections.
4. Inspect the fan, the fan and motor hangers, adjusting mechanism and mounting for any sign of deterioration or corrosion. Replace any missing or worn parts and lubricate as necessary with particular attention to the shaft and bearings.
5. Inspect disconnect (local isolation switch and remote) master. Look for damaged, loose or missing parts or connections, frayed or discolored wiring, and insure the unit is properly marked.
6. Test operate unit. Take a meter reading and record.
7. Spot paint entire unit as necessary.
8. Replace cover using all fasteners (don't skip).
9. Clean up unit and work area.

COMMENTS: Here you should write out exactly what work and what parts were required to do this particular job for each individual fan unit.

ONE STEP AT A TIME

INSPECTION NO. 12

a Planned Maintenance Program

PREVENTIVE MAINTENANCE FOR AIR CONDITIONING UNDER WINDOW OR WALL MOUNTED UNITS

1. Change filters.
2. Clear drain.
3. If oil, lubricate motor and fan.
4. If belt driven, inspect belt tension and condition. Also, check sheaves for alignment and wear.
5. If direct driven, inspect fan shaft lockscrews.
6. Inspect squirrel cage for dirt or unbalance.
7. Inspect the electrical wiring and the disconnect.
8. Operate (change the thermostat setting) and insure that the unit follows command.
   a. If refrigerated unit, compressor should respond.
   b. If chilled water unit, the control valve should respond.
9. Test operate unit and record discharge air temperature.

NOTE: Annual for this unit should include back flushing of coil, and if in a contaminated or dirty area, high pressure washing of coils.

NOTE: If this unit has steam or hot water heating:
   a. Insure trap is working.
   b. Insure control valve, hand or auto, is functioning.
   c. Look and listen for leaks and corrosion.

FINAL NOTE: Insure all covers are in place and area is clean.

ONE STEP AT A TIME

INSPECTION NO. 13

a Planned Maintenance Program

INSPECTION LIST FOR HOT WATER HEATING

Convectors under windows with or without fan units.

Boiler or heat exchanger.

Expansion Tank.

Pressure Control.

Circulating Pump.

Air Venting System.

Controls.

Safety Valve.

ONE STEP AT A TIME

INSPECTION NO. 14

a Planned Maintenance Program

**PREVENTIVE MAINTENANCE INSPECTION FOR
RECIPROCATING COMPRESSOR**

1. Gauges
    a. Clean.
    b. Calibrate.
    c. Check for proper operation.
    d. Check for leaks.
    e. Tighten fittings.
2. Motor
    a. Check operating current.
    b. Start coil and contacts.
    c. Lubricate motor bearings.
    d. Check motor insulation resistance.
3. Unit
    a. Measure ventilation.
    b. Check alignment of drive section.
4. Control Panel
    a. Calibrate and clean controllers and safety controls.
    b. Check setpoint or controls and limits.
    c. Sequence test all controls.
5. Compressor
    a. Check crankcase heater operation.
    b. Check refrigerant charge.
    c. Check for refrigerant and oil leaks.
    d. Test for efficiency.
    e. Check oil level and condition.
    f. Perform acid test.
    g. Observe bearing and operating surface temperature.
    h. Measure vibration.
    i. Sequence cylinder unloaders.
    j. Inspect head pressure safety valve.

ONE STEP AT A TIME

INSPECTION NO. 15

a Planned Maintenance Program

**PREVENTIVE MAINTENANCE INSPECTION FOR
REFRIGERATION-FREEZER COMPRESSOR**

1. Check all electrical wiring and connections, motor and safety devices.
2. Motor: Check for heat, noise, lubrication, sparking and shaft end play.
3. Compressor:
   a. Check for noise.
   b. Check amount of refrigerant. Is refrigerant dry?
   c. Is type of refrigerant used plainly marked on unit?
   d. Check oil level.
   e. Check water flow.
   f. Check condition of coil.
   g. Check for any gas leak.
   h. Check support for tubes and coils.
   i. Check entire area around unit for cleanliness and obstructions.
4. For a conventional unit, check:
   a. Belt condition.
   b. Belt alignment.
   c. Belt tightness.
5. For a hermatic condensing unit, check:
   a. Overload cutout relay.
   b. Capacitor.

ONE STEP AT A TIME

INSPECTION NO. 16

a Planned Maintenance Program

REFRIGERATION INSPECTION

1. Electrical connections	CHECKLIST
2. Oil level	1. Test for leaks
3. Water flow	2. Check refrigerant charge: the
4. Amount of refrigerant	head pressure and low-side
5. Dryness of refrigerant	pressure
6. Motor and safety devices	3. Check oil charge
7. Compressor noises	4. Clean evaporator
8. Gas leaks	5. Straighten fins
9. Coil conditions	6. Check and lubricate motor
10. Supports for tubing	7. Check belt condition and
11. Coil supports	tension
12. Cleanliness	8. Check water valve
13. For conventional condensing	9. Check water drain
unit, check condition of belt	10. Check circulating fans
its tightness and alignment	11. Voltage reading
14. For hermatic condensing unit,	12. Wattage reading
check the overload cutout,	ALSO
relay and capacitors	13. Label with refrigerant used
	14. Be sure insulation is on suction
	line
	15. Clean lens of refrigerant sight
	glass
	16. Oil filters

TYPES OF COMPRESSORS:
Compression/absorption,
Positive displacement, Reciprocating,
Rotary, Impeller, Centrifugal.

ONE STEP AT A TIME

INSPECTION NO. 17

a Planned Maintenance Program

PREVENTIVE MAINTENANCE MONTHLY INSPECTION FOR AIR CONDITIONING UNITS OTHER THAN PORTABLE

1. Change filters.
2. Clear drain.
3. If oil, lubricate motor and fan.
4. If belt driven, inspect belt condition and tension. Also check sheaves for alignment and wear.
5. If direct driven, inspect fan shaft lockscrews.
6. Inspect squirrel cage for dirt or unbalance.
7. Inspect the electrical wiring and the disconnect.
8. Test operate unit and record discharge air temperature.

NOTE 1: If this unit has steam or hot water heating:
   a. Insure trap is working.
   b. Look and listen for leaks and corrosion.

NOTE 2: Annual for this unit (try to schedule for month of May) "In addition to normal Monthly"
   a. Vacuum coils and inside of housing.
   b. Blow out drain.
   c. Back flush coil.
   d. Spot paint as necessary.
   e. In a contaminated or dirty air, high pressure wash coils.
   f. Operate (change the thermostat setting) and insure that the unit follows command.
      (1) If refrigerated unit, compressor should respond (temp-change)
      (2) If chilled water unit, the control valve should respond (temp-change)
      (3) If steam or hot water heating, control valve should respond (temp-change)
      (4) Test steam trap
      (5) Inspect all tubing (piping)

FINAL NOTE: Insure all covers are in place (in good repair) and area is clean.

ONE STEP AT A TIME

INSPECTION NO. 18

a Planned Maintenance Program

PREVENTIVE MAINTENANCE MONTHLY INSPECTION FOR AIR CONDITIONING—CEILING MOUNTED CHILLED WATER UNITS

1. Change filters.
2. Clear drain.
3. If oil, lubricate motor and fan.
4. If belt driven, inspect belt tension and condition. Also check sheaves for alignment and wear.
5. If direct driven, inspect fan shaft lockscrews.
6. Inspect squirrel cage for dirt or unbalance.
7. Inspect the electrical wiring and the disconnect.
8. Test operate unit and record discharge air temperature.

NOTE: Annual for this unit (try to schedule for month of May) "In addition to normal monthly"
   a.   Vacuum coils and inside of housing.
   b.   Blow out drain.
   c.   Back flush coil.
   d.   Spot paint as necessary.
   e.   In a contaminated or dirty air, high pressure wash coils.
   f.   Operate (change the thermostat setting) and insure that the unit follows command:
       (1) If refrigerated unit, compressor should respond.
       (2) If chilled water unit, the control valve should respond.

ONE STEP AT A TIME

INSPECTION NO. 19

a Planned Maintenance Program

CONVEYOR INSPECTION

The drive unit: gearbox, motor coupling, chain, belt, sprockets, drive shaft, etc.

The drive control: all electrical contacts, sensors, switches, lights, mountings, etc. This applies equally to belt, bucket, chain, roller and screw-type units. Inspect each roller, belt, screw, bucket, chain for any signs of wear and insure all connecting hardware is in place and in good repair. The crowned sheaves (rolls) on belt conveyors are the main alignment for the belt, so insure their good state of repair.

Inspect all tensioning and alignment items (include lacing, etc.). Inspect the belt cleaners, brushes, air nozzles and tubing; steam spray cleaning units and all drain systems.

Inspect the drive connecting devices, sprockets, chains, belts, couplings, pins, etc.

Inspect any auto lubricating systems to insure correct operation.

Inspect any auto fire suppression systems to insure their good state of repair.

Inspect the entire superstructure and the housing for any signs of deterioration.

Inspect and clear the area around the unit and inspect all guard systems around the unit.

ONE STEP AT A TIME

INSPECTION NO. 20

a Planned Maintenance Program

PREVENTATIVE MAINTENANCE FORM

MONTHLY ROUNDS			INSPECT								OK
DATE ISSUED_____											
COMPLETED											
I.D. NO.	NAME	LOCATION									
ONE STEP AT A TIME											

INSPECTOR

FORM 25

a Planned Maintenance Program

PREVENTATIVE MAINTENANCE FORM

MONTHLY ROUNDS			INSPECT								
DATE ISSUED_____ COMPLETED			UNUSUAL SOUNDS	VISUAL DETERIORATION	INSPECT FILTER	CHANGE FILTER		TIME IN MINUTES		DRAWING REFERENCE	OK
I.D. NO.	NAME	LOCATION									

ONE STEP AT A TIME

INSPECTOR

FORM 26

a Planned Maintenance Program

PREVENTATIVE   MAINTENANCE   FORM

MONTHLY ROUNDS			INSPECT									OK
DATE ISSUED_____ COMPLETED			UNUSUAL SOUND	VISUAL DETERIORATION	PUMP & MOUNTING	MOTOR & MOUNTING	COUPLING & GUARD	SEAL &/OR PACKING	PIPING/VALVES HANGERS, ETC.	ELECTRICAL WIRE/CONDUIT	PRESSURE IN/OUT	TIME IN MINUTES	
I.D. NO.	NAME	LOCATION											

ONE STEP AT A TIME

INSPECTOR

FORM 27

a Planned Maintenance Program

PREVENTATIVE MAINTENANCE FORM

MONTHLY ROUNDS			INSPECT										OK
DATE ISSUED_____ COMPLETED _____			UNUSUAL SOUND	PREFILTERS	PRICIPITATOR ON	ALL DRAINS	MOTOR & FAN BEARINGS	BELTS & COUPLINGS	READING ON MAGNEHELIX	ADD CHEMICALS AS NEEDED	PUMP-BEARINGS & PALK'G & SEALS	INDICATORS	
I.D. NO.	NAME	LOCATION											

ONE STEP AT A TIME

INSPECTOR

FORM 28

a Planned Maintenance Program

PREVENTATIVE   MAINTENANCE   FORM

MONTHLY ROUNDS

INSPECT

DATE ISSUED _____
COMPLETED

I.D. NO.	NAME	LOCATION	UNUSUAL SOUNDS	VISUAL DETERIORATION	FILTERS AIR & CONDENSATE	OIL LEVEL & PRESSURE	WATER TEMP. & PRESSURE	MOTOR BEARINGS COUPLING/BELT	COMPRESSOR PACKING/SEAL	AIR/VAC PRESSURE	CONTROLS & DISCONNECT	TIME IN MINUTES	OK

ONE STEP AT A TIME

INSPECTOR

FORM 29

C
Appendix C:
Forms to Accompany Step Seven

a Planned Maintenance Program

EXAMPLE ONLY

MAINTENANCE

WEEK NUMBER

NUMBER	1	2	3	4	5	6	7	8	9	10	11	12	13	14	15	16	17	18	19	20	21	22	23	24	25	26	27	28	29	30	31	32	33

FORM 30

ONE STEP AT A TIME

a Planned Maintenance Program

SCHEDULING

WEEK_____INDEX

52 WEEK FILE

Identity Number	Section Assigned	FREQUENCY						Remarks
		Wkly	Bi-Wkly	Mthly	Qtry	Semi-Annual	Annual	

ONE STEP AT A TIME

FORM 31

a Planned Maintenance Program

					MAINTENANCE		
ITEM	P.M. ID NO.	BLDG NO.	LOCATION	P.M. DWG REF	SECT	SCHL	PRIORITY

SCHEDULING INFORMATION
IDENTIFICATION NUMBER ASSIGNMENT

ONE STEP AT A TIME

FORM 32

a Planned Maintenance Program

EXAMPLE ONLY

TIME ALLOCATION _____ DAY _____ DATE _____

										MAINTENANCE					

Equipment Information			Action Indicator			**MAINTENANCE**						

Category	Type	Sequence Number	Frequency	Priority	Function	Estimated Time Required	TIME					
							Carpenter		Electronic		Electrical	
							Asgnd	Avail	Asgnd	Avail	Asgnd	Avail

ONE STEP AT A TIME

FORM 33

a Planned Maintenance Program

PAGE _____

Mechanical		Painter		Plumber		Pneumatic		Refrigeration		Groundsman		Overtime	
Asgnd	Avail	Asgnd	Avail	Asgnd	Avail	Asgnd	Avail	Asgnd	Avail	Asgnd	Avail	Asgnd	Auth

ONE STEP AT A TIME

FORM 33

a Planned Maintenance Program

DEFERRED SCHEDULING						
ISSUE				Deferred (Why)	Reissue P.M. Work	Comments
P.M. Work	ID No.	Maint Sect.	Completed Date			
ONE STEP AT A TIME						

FORM 34

D
Appendix D:
Forms to Accompany Step Nine

a Planned Maintenance Program

MONTHLY ACTION REPORT

BLDG NO. _____

NAME: _____

MONTH _____

Date		P.M. ID No.	Priority	Item Description	Parts Needed/Used	Time in Minutes				Type Inspection	Material Cost Indicated
Reported	Completed					Elec.	HVAC	Mech.	Plum.	Other		

ONE STEP AT A TIME

FORM 35

a Planned Maintenance Program

PLANNED MAINTENANCE ACTION REPORT

Month of _____   Date _____

A. Total new actions reported _____
    Total actions completed _____

B. Total manhours reported _____
    By sections _____
                         Elec.–HVAC–Mech.–Plumb.–Other
    1. Daily rounds _____
    2. Monthly rounds _____
    3. PM action/Job orders _____
    4. Water Treatment _____
    5. Other (   ) _____

    Total each section

C. Total material cost reported _____
    (See Action Report Forms for breakdown)

D. Current Action:
    1. _____
        _____
        _____

    2. _____
        _____

SPECIAL NOTE: _____
_____
_____
_____
_____

ONE STEP AT A TIME

FORM 35-A

a Planned Maintenance Program

PLANNED MAINTENANCE REQUEST

Date _____

1. Originator _____

   Name   Title   Dept.

2. Item _____

   Name   Description   Serial No.

   _____

   Manufacturer   Date Installed   Date in Service

3. Location _____

   Building   Room

4. Use _____

   Describe

   _____

5. Publications _____

   Operation

   _____

   Maintenance

6. Any Special Requirements _____

   _____

7. Acceptance _____

   (Yes) (No)   (Reason)

   _____

8. Planned Maintenance ID No. _____

   Assigned   Date

Comments:

ONE STEP AT A TIME

FORM 36

a Planned Maintenance Program

SAFETY
Maintenance / Engineering / Department
INCIDENT REPORT

_____ Action Requested   Date _____

_____ Info Only

WHEN

WHAT

WHERE

HOW

WHY

INJURY
Employee Name_____ Number_____

Sent   Hospital ☐   Home ☐   Work ☐

Equipment Lost

_____   _____
                Supervisor   Section

ONE STEP AT A TIME

FORM 37

a Planned Maintenance Program

SAFETY
Maintenance / Engineering / Department
INCIDENT REPORT

SUMMARY

Date _____
Period   From–To

UNSAFE   TOTAL

  Act   _____

  Condition   _____

  Equipment   _____

  Tool   _____

INJURY _____

EQUIPMENT LOST _____

UNAVOIDABLE _____

SECTION TOTAL

_____
Prepared By

ONE STEP AT A TIME

FORM 38

a Planned Maintenance Program

VEHICLE TRIP RECORD

VEHICLE ID NO.————————————— DATE————————

NOTE:

EACH OPERATOR CHECK:

Brakes, Lights, Horn, Tires, Steering, Backup Warning Device,
General Overall Condition, Gas, Oil, Radiator Water, Battery Water,
and Windshield Washer Fluid.

Record any GAS ——————————— added.

OIL ———————————

WATER ———————————

Do Not Operate This Vehicle If Any Repairs Are Needed.
Notify Supervisor Immediately.

Operator	Mileage Out	Mileage In	Destination	Time of Depart.	Time of Arrival	Number of Passengers

THIS FORM TO BE ON CLIPBOARD IN ASSIGNED VEHICLE
WHEN VEHICLE IS BEING OPERATED.

ONE STEP AT A TIME

FORM 39

a Planned Maintenance Program

Preventive Maintenance
ACCEPTANCE CHECKLIST

DATE _____

PERSON INSPECTING
CONTRACTOR/REP   OWNER/REP.

ITEM DESCRIPTION   MODEL _____

SERIAL NO. _____   MAINT. ID NO. _____

LOCATION

ITEM IS INSTALLED AS INDICATED ON DRAWING NO.

ITEM IS COMPLETE AS INDICATED BY PUBLICATION TITLE AND DATE

ITEM IS PROPERLY GUARDED

ITEM HAS BEEN TEST OPERATED BY CONTRACTOR

INDICATORS ARE LIGHTS, GAGES, SOUND, ETC.   WHAT IS INDICATED

LIST ANY SPECIAL TOOLS/TEST/ETC. THIS ITEM REQUIRES

LIST ANY NOTABLE OPERATION OR MAINTENANCE PROBLEMS

PREOPERATION

ALL LUBRICATION HAS BEEN COMPLETED
ALL UTILITIES/FUEL ARE PROPERLY SUPPLIED AND CONNECTED
ALL NECESSARY DRAINS ARE OPEN
GUARDS ARE IN PLACE
ALL SAFETY PRECAUTIONS ARE COMPLETED
MAIN DISCONNECT/VALVES ARE ON
UNIT SHOWS NO SIGN OF DETERIORATION

OPERATION

STARTS
OPERATES CORRECTLY
MALFUNCTIONS NOTED: HOT BEARINGS, NOISY, LEAKS, STOPS, ETC.
ALL CONTROLS FUNCTION PROPERLY
ALL VALVES OPERATED: NO LEAKS AIR/FLUID
INDICATORS FUNCTIONING PROPERLY: LIST CORRECT READINGS
OR SETTINGS

AFTER OPERATION

INSPECT   A.   FUEL   F.   MOUNTING/HARDWARE
          B.   OIL   G.   VALVES/TRAPS/STRAINERS/
          C.   LUBRICANT   PIPING
          D.   FILTERS   H.   OVERALL FOR ANY SIGN OF
          E.   BELTS/COUPLINGS   DEFICIENCY OR
                                        DETERIORATION

ONE STEP AT A TIME

FORM 40

E

Appendix E:
Enclosures to Accompany
Step Ten

a Planned Maintenance Program

ENERGY INFORMATION

The Community and Energy Savings

All users and workers must begin to pay careful attention to the things we have for so long taken for granted. Are the lights necessary? Is the heat too high? Is the air conditioning too cold? Is the air in your area blowing paper, holding doors open? Do all the doors work and are they properly closed, or are they held opened for some unauthorized reason? Do you have a window cracked open, or one that won't close? Do you stack paper, books, other items on air registers (look in front of window)? Do you have incandescent lights in the ceiling fixtures?

In your area have you any high energy users (heaters, stoves, toasters, hot plates) or special portable equipment that you use in your work?

Any of the above items found should be corrected and items that require engineering to work on should be reported to the energy control office.

Please check your thermostat setting.

Turn out lights when you don't need them and especially when you leave the area.

These items will be checked, and items reported that are not done or reported will cause a citation to be issued.

Items brought to our attention will also receive an award.

Did you use an elevator when you could have walked? 1 flight, 2 flights?

ONE STEP AT A TIME

ENCLOSURE NO. 1

a Planned Maintenance Program

SECURITY AND ENERGY SAVINGS

On exterior inspection of the facility, report to the energy office any doors or windows found to be open, loose hung, parts missing or in any manner allowing air to enter or exhaust from the building.

Report any lights found on (unauthorized), or note ones you feel are unnecessary.

Any water found running should be reported to the energy control office.

On patrol throughout the facility any sounds of hissing steam should be reported to the energy control office.

Any complaints on lighting, heating and cooling should be reported to the energy control office.

ONE STEP AT A TIME

ENCLOSURE NO. 2

a Planned Maintenance Program

HOUSEKEEPING AND ENERGY SAVINGS

Check for windows or doors that won't quite close, or windows that you can feel air coming in around. Look for cracked, broken or missing panes or panels, etc., exterior doors with openings showing above or below or by the hinges, etc., and report to the energy control center for correction. Of course you should always close any window or door left open and report that to the energy control office. (Any exterior door found open should be reported to Security immediately.)

Any sink faucet found running should be turned off, and if it continues to drip, make a special note on report to energy control office.

Any lights that are not for security or fire protection should be turned off and any found on should be reported to energy control office.

Any area you enter that is overheated or cold should be reported to maintenance immediately and a report sent to the energy control office.

In all areas where there are air vents in walls, ceilings, etc., including offices, rooms, hallways, janitor closets, the vents should be kept clear of obstruction and the grills themselves should be kept clean by brushing. Any obstructed ones should be reported to the energy control office. In many of the offices there are air vents at the windows and people tend to stack books, papers and other material over the vents. These should be kept clear and any vents found covered should be reported to the energy control office. NOTE: Do not move these items as they may be in special order or there may be some other reason for not moving them.

Report any small heaters (portable units) found in any areas to the energy control office. (Also, coffee pots and hot plates.)

In the wintertime in the evening, insure that all window blinds, drapes and other items intended for closing are closed, drawn, or in place.

Should you feel the water in the sinks is hotter than normal, report that to the energy control office as well.

ONE STEP AT A TIME

ENCLOSURE NO. 3

Index